Taming the Flame: Advancing Clean Combustion Technologies

Nicholas

Contents

Chapter 1

Introduction

1.1 Motivation

Kein Grad weiter! (Not any further degree). This was the motto with which end of September the movement Fridays for Future came back to the scene after several months of public absence due to the Covid-19 crisis [14]. Climate change effects are already perceivable through alterations in the hydrological cycle [87], increased frequency, intensity and duration of heat-related events[21], crop yield reductions[63][60], sea level rise[1][62], ice melting and glacier mass loss[106][78], to cite only a few. The point at which global consumption exceeds its sustainable rate, i.e. the Earth Overshoot Day, takes place earlier each year[86]. In 2017 Hurricane Harvey severely hit the USA, with an impact of around 90billion USD, out of which between one third and three quarters of the damages could be attributable to the human influence on climate[54]. Though the cost estimation method might be arguable, the case illustrates the gravity of the situation, and together with the other mentioned facts, it highlights the urgency for measures to be taken.

The Paris Agreement aims to keep the global temperature rise this century well below $2°C$ and to pursue efforts to limit the temperature increase even further to $1.5°C$ [139]. This should be done at the same time that future energy demand is likely to increase not only due to socioeconomic developments, but as well due to climate change[143]. Complexity is added to the problem by acknowledging the energy transition multidisciplinary character. For instance, due to the chemical substances and emissions involved in photovoltaic panels production, they cannot be considered as a zero-emission technology[76]. Moreover, aspects like the technology acceptance play a non-negligible role[50][80], e.g. wind energy projects are witnessing increasing levels of local opposition[31].

Combustion will continue to play a paramount role in the energy mix, e.g. through biomass [95][2][94], at the same time that fossil fuels will be progressively reduced, as they are the main cause for global CO_2 emissions [128][7].The field of research is wide, as new fuels, models and technologies should be developed to make the combustion process more clean and efficient. Facing such an overwhelming challenge, this work aims to contribute on the specific domain of turbulent combustion model development. More specifically, this research project lies within the computational fluid dynamics (CFD) framework. It presents a numerical strategy to simulate non-premixed turbulent combustion employing tabulated chemistry in Large Eddy Simulations (LES) context. The novel approach departs from well established premixed combustion concepts such as direct flamelet filtering, and the corresponding flame front wrinkling, and adapts them to fully exploit their capabilities in the non-premixed regime.

1.2   State of the art

Given the increasing computational power, the availability of Direct Numerical Simulations (DNS) of reacting flows with billion mesh points has substantially increased in the last years [3][19][24][99]. Such an approach offers advantages such as remarkably high resolution, data accessibility and assessment at conditions where experiments are difficult to conduct[127]. The data bases have been employed among others for a-priori model evaluation [152][90], flame front characterization[129] and for machine learning [45][46][79][149]. Though the size of the employed mechanisms has increased, e.g. 47 species and 290 reactions [4][5], chemistry reduction continues to be a requirement on DNS simulations. Moreover, this approach requires immense computational resources, for instance, $1e^8$ cpu-hours executing on up to 131,072 processors, were necessary to generate the data base in [88] which exceeds 230 TB in size. As a result DNS remain limited to few realizations at specific conditions, thus being not yet appropriate for systematic statistical analysis[69].

Large Eddy Simulation (LES) is a well developed approach[56] where only the larger scale motions are resolved on the numerical grid and the smaller scale motions are accounted for by a subgrid-scale (SGS) model[126]. Despite the substantial decrease with respect to DNS[89], LES remains computationally expensive, thus restricting the use of relatively small detailed chemical mechanisms. Due to unceasing discoveries in the field of chemical kinetics the size of detailed chemical mechanisms continuously augments[37], therefore as for DNS, in LES reduction techniques are unquestionably necessary in order to be able to simulate practical combustion problems.

Numerous methodologies and categorizations have been proposed, for instance into mechanism reduction, storage/retrieval-based methods, and adaptive chemistry strategies[81]. The latter exploits the fact that a specific composition space region may require fewer species and reactions than a more detailed model applicable over the entire composition space[83]. Powerful adaptive chemistry techniques have been developed considering both on-the-fly [82][61] and pre-processing strategies [132][133]. For instance the Sample Partitioning Adaptive Reduced Chemistry (SPARC) procedure has delivered promising results[47] combining Local Principal Component Analysis (LPCA)[74] and mechanism reduction using Directed Relation Graph with Error Propagation (DRGEP)[109].

Another type of approach consists to reduce the number of transported equations exploiting the fact that the energy and species equations can be expressed in a common form given some reasonable assumptions, e.g. equal species and thermal diffusivities, no body forces, and no pressure gradient diffusion[153]. Coupling functions, i.e. linear combinations of these equations, can then be derived to eliminate the source terms, thus resulting into conserved scalars. One of the first solutions of this type can be attributed to Burke and Schumann[15], whose flame sheet model assumed infinitely fast chemistry and constructed the coupling function as the sum of the fuel and the oxidizer mass fractions. Bilger further developed the conserved scalar formalism proposing the mixture fraction as the mass of atoms in a two streamed system, and demonstrated that under the equilibrium assumption all the scalars could be uniquely related to this mixture fraction[11]. The laminar flamelet model proposed by Peters[110] extended the conserved scalar approach to consider non-equilibrium effects by adding the instantaneous scalar dissipation rate at stoichiometric conditions and set a milestone on the development of flamelet approaches.

Flamelet based methods assume thin flame structures with significantly smaller time scales compared to the flow. Exploiting this feature, the chemistry description can be parametrized in terms of few controlling variables resulting in significant computational time reduction when using complex chemical-kinetics mechanisms[131][141]. Within this framework, FGM (Flamelet Generated Manifold)[142] and FPI (Flamelet Prolongation of ILDM)[57], both initially developed for premixed and afterwards extended to non-premixed regime, as well as FPV (Flamelet/Progress Variable)[114][65] method for non-premixed regime are consolidated techniques.

Flamelet tabulation has been applied for turbulent combustion both in RANS[91] [103], and in LES[102] [93] context. Refering to the SGS closure a well known approach consists to use presumed PDF to estimate the turbulence-chemistry interaction, though other strategies, e.g. conditional source term estimation[16] or artificial inteligence[48] have been explored as well. Within the presumed PDF framework, the approach has been further developed considering among others the number of parameterizing variables[154][151], the addition of unsteady flamelets [116][115], and the type of presumed functions[27]. Acknowledging the flexibility of flamelet methods, further developments have permitted the assessment of highly specialized applications, e.g., ethanol spray combustion [124], where FGM is coupled with an ATF (Artificially Thickened Flame)[29] approach employing dynamic flame surface wrinkling, or oxy-flame combustion by means of FPV coupling with stochastic fields methodology[92].

Tackling premixed turbulent combustion, the FTACLES (Filtered Tabulated Chemistry for LES) combustion model [53] is based on the direct application of a filtering operator to one-dimensional laminar flamelets. The strategy to estimate the closure terms ensures both the correct flame propagation as well as the recovery of the chemical structure of the filtered flame. The formalism has been extended to partially premixed regimes [6] and heat loss effects [97]. Currently, work has been carried out regarding non-premixed flamelet filtering into LES context. Two approaches have been proposed which differ in the way the flamelet modification through turbulence is taken into account. The approach followed in [150] and [20] derives the turbulent flamelet equation from the filtered species transport. Subsequently, similarity law and scaling relations in order to estimate the filtered chemical source term and scalar dissipation rate are proposed based on a-priori DNS study in [150], while [20] suggests a source term estimation employing first-order conditional moment closure.

Another way to address the problem is to directly filter the laminar flamelet and model the turbulence effect. Following this reasoning, the non-premixed filtered tabulated chemistry model as proposed by Coussement [32] has been designed to tackle turbulent combustion problems where the flame wrinkling is fully resolved or for laminar flame fronts. The model has been applied to 1-D and 2-D unresolved counterflow flame configurations and encouraging results have been reported. However, the model extension as to consider a more realistic configuration has not yet been done.

Among the wide range of conditions that can be found in turbulent combustion, the FTACLES concept specifically targets the description of an unwrinkled flame front which cannot be resolved by the numerical grid. For this very precise situation, the formalism offers an exact mathematical characterization of the modified flame structure resulting from a filtering operation. Focusing on the non-premixed regime, liquid rocket engines present a practical application that fulfills the above mentioned conditions, i.e. the flame front is very thin [0.1–0.01]mm and weakly wrinkled by turbulent motions [32]. The study of such systems is not only highly interesting from the numerical perspective but also quite relevant from the industrial point of view. Compared to the well established alternative of presumed PDF, on FTACLES the variance is replaced by the filter size in the tabulated parameterization, which results into an improvement from the computational point of view. Furthermore, the methodology directly introduces the effect of the numerical grid size into the combustion model, being thus highly consistent with the LES technique. A systematic assessment, validation and insightful scrutiny of the non-premixed FTACLES formalism is therefore highly necessary to employ it on real configurations. This issue motivates this contribution.

1.3 Purpose and outline

The final goal of this work is to assess the performance of the non-premixed FTACLES formalism on turbulent combustion. For this, a systematic approach is followed where first the numerical implementation is verified, afterwards validation is done on a coflow laminar diffusion flame, and finally the model full capabilities are assessed on the more complex turbulence condition. The structure of the work is as follows:

- Chapter 2 presents a non-exhaustive theory overview. It addresses on a first part the mathematical description of fluid flows and turbulence modeling. Subsequently a brief description of the numerical methods employed is presented. The last part is devoted to combustion theory, and it presents a selection of turbulent combustion models having some affinity with the non-premixed FTACLES approach.

- Chapter 3 is devoted to the model description. The flamelet equations are introduced, the filtering operator is applied, and the derivation of the non-premixed FTACLES model is presented. The model is appraised considering an individual flamelet and the entire manifold. The undergoing transformations are identified and extensively discussed. A comparison is performed between the non-premixed FTACLES formalism, the premixed FTACLES and the presumed PDF approach. Analogies are established and the specificities of the formalism are highlighted as well. The formalism is verified employing one and two-dimensional geometries.

- Chapter 4 elucidates the physical interpretation of the filtering operation by appraising the performance of the filtered tabulated chemistry model on a 3-D real laminar non-premixed coflow flame. The model results are compared against the direct filtering of the fully resolved laminar diffusion flame showing that the formalism adequately describes the underlying physics. The study reveals the importance of the one-dimensional counterflow flamelet hypothesis, so that the model activation under this condition is ensured by means of a flame sensor. The consistent coupling between the model and the flame sensor adequately retrieves the flame lift-off and satisfactorily predicts the profile extension due to the filtering process.

- Chapter 5 presents the first appraisal of the non-premixed FTACLES model under turbulent conditions. The results demonstrate that the formalism coupling with a SGS modeling function can adequately describe a wrinkled flame front condition. The model performs significantly well employing a three-dimensional tabulation strategy, where the numerical grid is coupled with the model by the third parameter, i.e. the computational cell size A secondary test, with a test-filter-like approach and two dimensional parameterization in terms of Z and c, shows a good model response for coarse numerical grids. The obtained results confirm the idea that SGS closure in diffusive combustion can be derived based on filtering arguments, and not only based on statistical approaches.

- Chapter 6 presents a summary of the work and proposes an outlook.

Chapter 2

Theoretical Background

2.1 Mathematical modeling and description of turbulent flows

2.1.1 Governing Equations

A convenient strategy when studying fluid flows consists to evaluate the behavior within fixed spatial regions, rather than to follow a parcel of matter in space, as done when analyzing solid bodies dynamics. Thus the control volume CV is defined as the region Ω, bounded by a closed surface S, and φ is an intensive property, i.e. a property per unit volume. Knowing that the flow can transport φ into or out of the domain, the flux $\mathbf{F}$ will be then defined as the amount of φ crossing the unit of surface per unit of time. The conservation law states that the variation in time of the quantity φ contained inside the domain Ω is equal to the flux of φ all along S, plus the contributions of sources that either generate or consume φ.

$$\frac{\partial}{\partial t} \int_{\Omega} \varphi d\Omega = - \oint_{S} \mathbf{F} \cdot d\mathbf{S} + \int_{\Omega} Q_V d\Omega + \oint_{S} \mathbf{Q_S} d\mathbf{S}, \tag{2.1}$$

where $\mathbf{F}$ is the flux vector, $\mathbf{S}$ the surface element with its vector pointing in the outward normal direction, and Q_V and Q_S the volumetric and surface sources respectively. The fact that the flux term is numerically expressed as the scalar product between $\mathbf{F}$ and $\mathbf{S}$ reflects that only the flux component normal to the surface can contribute to the change in φ, while the minus symbol makes the flow entering the domain to have a positive contribution. Employing Gauss theorem Equation 2.1 can be expressed in differential way

$$\frac{\partial \varphi}{\partial t} + \frac{\partial (F_j - Q_{s,j})}{\partial x_j} = Q_V. \tag{2.2}$$

Fluid flows can be fully described considering the conservation of three quantities, namely, mass, velocity and energy.

Mass conservation

Considering now the mass as the observed property, namely $\varphi = \rho$, and recalling that the mass inside the domain does not change, then the continuity equation is obtained

$$\frac{\partial \rho}{\partial t} + \frac{\partial \rho u_i}{\partial x_i} = 0, \tag{2.3}$$

Momentum conservation

Assuming a newtonian fluid it can be written as

$$\frac{\partial \rho u_j}{\partial t} + \frac{\partial \rho u_i u_j}{\partial x_i} = \frac{\partial \tau_{ij}}{\partial x_i} - \frac{\partial p}{\partial x_j} + F_j \,, \tag{2.4}$$

where F_j is a body force, τ_{ij} is the viscous stress tensor

$$\tau_{ij} = 2\mu S_{ij} - \frac{2}{3}\mu \delta_{ij} \frac{\partial u_j}{\partial x_j} \,, \tag{2.5}$$

and S_{ij} is the rate of strain tensor.

$$S_{ij} = \frac{1}{2}\left(\frac{\partial u_i}{\partial x_j} + \frac{\partial u_j}{\partial x_i} \right) \,. \tag{2.6}$$

Species transport

The species conservation can be written as

$$\frac{\partial \rho Y_k}{\partial t} + \frac{\partial \rho u_i Y_k}{\partial x_i} = -\frac{\partial \rho V_{i,k} Y_k}{\partial x_i} + \dot{\omega}_k \,, \tag{2.7}$$

where u_i is the velocity of the fluid mixture and $V_{i,k}$ the diffusion velocity of species k. Considering Fick's law

$$\frac{\partial \rho Y_k}{\partial t} + \frac{\partial \rho u_i Y_k}{\partial x_i} = \frac{\partial}{\partial x_i}\left(\rho D_k \frac{\partial Y_k}{\partial x_i} \right) + \dot{\omega}_k \,, \tag{2.8}$$

where D_k is the diffusion coefficient of species i into the mixture. If equal diffusivity is assumed for all species $D_k = D$, Equation 2.8 guarantees continuity, otherwise a correction velocity V_i^c as proposed by [28] should be introduced

$$V_i^c = \sum_{i=1}^{k} D_k \frac{\partial Y_k}{\partial x_i} \,, \tag{2.9}$$

resulting

$$\frac{\partial \rho Y_k}{\partial t} + \frac{\partial \rho (u_i + V_i^c) Y_k}{\partial x_i} = \frac{\partial}{\partial x_i}\left(\rho D_k \frac{\partial Y_k}{\partial x_i} \right) + \dot{\omega}_k \,. \tag{2.10}$$

Energy transport

The energy conservation can be expressed in several ways, depending on the variable, e.g. internal energy, enthalpy, and as well considering the total or sensible property. The convenience of a particular expression will depend on the specific problem to be solved. For instance enthalpy h conservation can be expressed as

$$\frac{\partial \rho h}{\partial t} + \frac{\partial \rho u_i h}{\partial x_i} = \frac{Dp}{Dt} - \frac{\partial q_i}{\partial x_i} + \tau_{ij}\frac{\partial u_i}{\partial x_j} + \dot{Q} + \rho \sum_{k=1}^{N} Y_k f_{k,i} V_{k,i} \,. \tag{2.11}$$

where $\frac{DP}{Dt}$ is the substantial derivative of the pressure, q_i the energy flux, $\tau_{ij}\frac{\partial u_i}{\partial x_j}$ is the viscous heating source term, $\dot{Q}$ the heat source term and the last LHS term is related to volume forces f_k on species k.

2.1.2 Turbulence

In order to appraise turbulent problems different strategies have been developed
which differ among each other on the extent to which the turbulence structures
are resolved. On the DNS approach no modeling is carried out and all the turbulence scales are explicitly determined. The RANS methodology is located on the
opposite extreme of the spectrum, it considers the averaging of the instantaneous
balance equations, thus solving uniquely for the mean values. Closure models are
employed and turbulence effects are then fully modeled. The LES technique resolves
the instantaneous filtered balance equations, the larger scale motions are explicitly
resolved on the numerical grid and the smaller scale motions are accounted for by
a subgrid-scale (SGS) model[126]. Figure 2.1 illustrates the model classification
considering the energy spectrum.

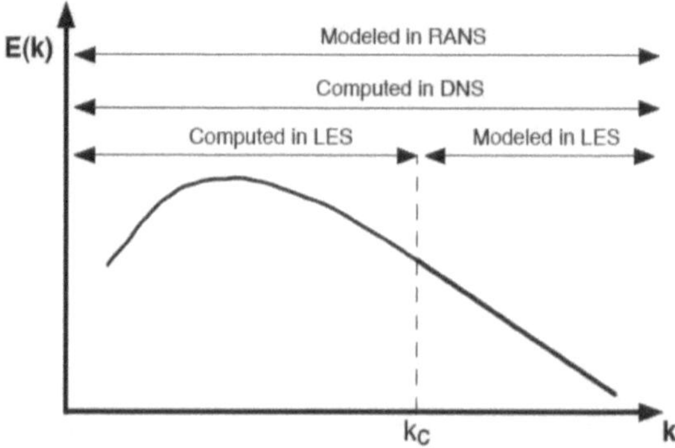

Figure 2.1: Turbulence energy spectrum as a function of the wave number with
RANS, LES and DNS frequency range. k_c is the cut-off wave number in LES. [120]

Table 2.1 summarizes the main features of the different turbulence modeling
approaches. The last column presents the required computational time, which is
indeed a fundamental feature as it defines the range of applicability of each technique. DNS requires immense computational resources[88] and as a result it remains
limited to few realizations at specific conditions, thus being not yet appropriate for
systematic statistical analysis[69]. Large Eddy Simulation (LES) is a well developed
approach[56], with a substantial decrease with respect to DNS[89]. Nonetheless LES
continues to be computationally expensive and restrictions to its application should
be considered, i.e. the use of relatively small detailed chemical mechanisms. Due to
its low computational demand RANS still prevails as an attractive alternative for
instance in the industrial context.

For the sake of clarity the expressions will be derived for the case of an incom-
pressible flow. The same principles hold for the compressible condition, where the
filtering/averaging operations should be simply expressed in terms of their Favre-
averaged counterparts.

Table 2.1: Turbulence modeling classification. The last column illustrates state of the art values of numerical grid sizes, and does not correspond to any fixed criterion.

Approach	Solved Structures	Modeled	Variable	Comput. cost	Grid points
DNS	All	None	Instantaneous	Very high	$1e^7 - 1e^9$
LES	Biggest	Subgrid	Instantaneous	Interm. / High	$1e^6 - 1e^7$
RANS	None	All	Mean	Low	$< 1e^6$

RANS

The Reynolds averaged decomposition expresses a variable $u(x, t)$ as:

$$u = \langle u \rangle + u'$$

(2.12)

where $\langle u \rangle$ is the time averaged value, and can as well be interpreted as the ensemble average over a given number of repetitions, and u' are the fluctuations around this mean.

Considering that $\langle u' \rangle = 0$, $\langle \langle u \rangle u' \rangle = 0$, and $\langle \langle u \rangle \rangle = \langle u \rangle$, the convective term of equation 2.4 can be expressed as

$$\langle u_j u_i \rangle = \langle u_j \rangle \langle u_i \rangle + \langle u'_j u'_i \rangle \,,$$

(2.13)

where the second RHS term corresponds to the flux caused by the fluctuating or turbulent part, and originates the Reynolds stress tensor defined as

$$\tau_{ij}^R = \langle u'_j u'_i \rangle \,.$$

(2.14)

The Reynolds averaged momentum equation is obtained by applying this decomposition to equation 2.4

$$\frac{\partial \langle u_i \rangle}{\partial t} + \frac{\partial \langle u_j \rangle \langle u_i \rangle}{\partial x_j} = \frac{\partial \langle \tau_{ij} \rangle}{\partial x_j} - \frac{\partial \tau_{ij}^R}{\partial x_j} - \frac{1}{\rho} \frac{\partial \langle p \rangle}{\partial x_i} \,.$$

(2.15)

Since τ_{ij}^R is unknown, a closure should be made, for which an natural idea is to derive equations for the Reynolds stress, originating the so-called Reynolds stress models. Another widely employed closure alternative is the eddy viscosity hypothesis, which relates the deviatoric part of τ_{ij}^R to the resolved $\langle \tau_{ij} \rangle$ by means of the turbulent viscosity ν_t as

$$\tau_{ij}^R - \frac{2}{3} k \delta_{ij} = -2\nu_t \langle S_{ij} \rangle \,,$$

(2.16)

where k is the turbulent kinetic energy

$$k = \frac{1}{2} \langle u'_i u'_i \rangle = \frac{1}{2} tr(\tau_{ij}^R) \,.$$

(2.17)

The isotropic component being added to the pressure, and so resulting into a modified pressure

$$p^{mod} = \langle p \rangle + \frac{2}{3} \rho k \delta_{ij} \,.$$

(2.18)

The final momentum equation resulting

$$\frac{\partial \langle u_i \rangle}{\partial t} + \frac{\partial \langle u_j \rangle \langle u_i \rangle}{\partial x_j} = \frac{\partial}{\partial x_j}\left(2\nu_{eff}\langle S_{ij}\rangle\right) - \frac{1}{\rho}\frac{\partial p^{mod}}{\partial x_i}. \tag{2.19}$$

Considering the decomposition of any given arbitrary transported scalar φ, equation 2.13 can be rewritten as

$$\langle u_j \varphi \rangle = \langle u_j \rangle \langle \varphi \rangle + \langle u_j' \varphi' \rangle. \tag{2.20}$$

The second RHS term gives origin to the turbulent scalar flux $\rho \langle u_j' \varphi' \rangle$, and its closure can be done by means of a gradient diffusion hypothesis

$$\langle u_j' \varphi' \rangle = -D_t \frac{\partial \langle \varphi \rangle}{\partial x_j}, \tag{2.21}$$

where D_t is the turbulent diffusion, and it is generally computed based on the turbulent viscosity and an adimensional number as Sc or Pr, depending on the considered scalar φ. This term is finally added to the molecular counterpart as $D_{eff} = D + D_t$, and the transport equation results

$$\frac{\partial \langle \varphi \rangle}{\partial t} + \frac{\partial \langle u_j \rangle \langle \varphi \rangle}{\partial x_j} = \frac{\partial}{\partial x_j}\left(D_{eff}\frac{\partial \langle \varphi \rangle}{\partial x_j}\right) - \langle \omega_\varphi \rangle. \tag{2.22}$$

Closure models

The turbulent viscosity is unknown and closure models should be then employed as to be able to solve the averaged transport equations. Following a dimensional analysis it can be seen that the dynamic turbulent viscosity μ_t can be expressed in terms of a velocity scale v and a lengthscale L as

$$\mu_t = C_\mu \rho v L, \tag{2.23}$$

where C_μ is a dimensionless constant. The mixing length model for instance estimates $v = L\frac{\partial \langle u \rangle}{\partial y}$ and the model calibration enters through the definition of L. Such an approach is too simplistic, and a significant improvement is obtained by transporting the turbulent kinetic energy s.t. $v = 2k^{1/2}$ as in the one equation model where

$$\mu_t = C_\mu \rho k^{1/2} L. \tag{2.24}$$

Two equation models overcome the inconvenience of flow specification through L by transporting a second parameter. A quite popular option is the $k - \varepsilon$ formalism, that exploits the fact that in equilibrium turbulent flows, where production and dissipation balance, the relation $L = \frac{k^{3/2}}{\varepsilon}$ holds, and then

$$\mu_t = C_\mu \rho \frac{k^2}{\varepsilon}. \tag{2.25}$$

Due to the complexity of the exact expression, the employed transport equation for the turbulent energy dissipation ε is rather empirical, and the model employs five constants, for which the most common values are: $C_\mu = 0.09$, $C_{\varepsilon 1} = 1.44$, $C_{\varepsilon 2} = 1.92$, $\sigma_k = 1.0$ and $\sigma_\varepsilon = 1.3$. Another well known two equation model is $k - \omega$, where the second transported variable ω is an inverse time scale.

LES

On Large Eddy Simulations the biggest turbulent structures, which are responsible for most of the transport of the conserved properties are solved, while the smallest ones are modeled. Contrary to RANS this is not a time but a spatial averaging operation where the filtering takes the form:

$$\overline{\varphi}(x) = \int G(x, x')\varphi(x')dx' \tag{2.26}$$

where $G(x, x')$ is the filter function, e.g. Gaussian, box filter or cutoff. In this way the variable $u(x, t)$ will now be decomposed as:

$$u = \bar{u} + u^{SGS}. \tag{2.27}$$

where $\bar{u}$ is the resolved or filtered quantity and u_{SGS} corresponds to the unresolved or subgrid component. Applying it to equation 2.4 the expression results

$$\frac{\partial \bar{u}_i}{\partial t} + \frac{\partial \bar{u}_i \bar{u}_j}{\partial x_j} = \frac{\partial \overline{\tau}_{ij}}{\partial x_j} - \frac{\partial \tau_{SGS}}{\partial x_j} - \frac{1}{\rho}\frac{\partial \bar{p}}{\partial x_i}, \tag{2.28}$$

where τ_{SGS} is the subgrid scale stress tensor. It corresponds to the difference between the exact filtered convective term and the employed approximation

$$\tau_{SGS} = \left(\overline{u_i u_j} - \bar{u}_i \bar{u}_j\right). \tag{2.29}$$

Due to the filter properties, the combined terms emerging from equation 2.27 do not fall away, τ_{SGS} does not only contain the Reynolds contribution but other terms are included as well and consequently $\tau_{SGS} \neq \overline{u_i^{SGS} u_j^{SGS}}$.
A well known closure model is the one proposed by Smagorinsky, which employs an eddy viscosity assumption. The turbulent viscosity is defined in terms of the filter size Δ and the characteristic filtered rate of strain $\bar{S} = (\bar{S}_{ij}\bar{S}_{ij})^{1/2}$ as

$$\mu_t = C_S^2 \rho \Delta^2 \bar{S}, \tag{2.30}$$

where C_S is a model constant, generally used as $C_S = 0.2$, but this might change as function of Re and flow conditions. Moreover wall functions should be introduced as to damp μ_t since the model is not able to adequately handle their effect.
An alternative approach is to exploit the similarity between the smaller resolved scales and the larger modeled ones, emphasizing that the most important interaction between both scales occurs around Δ. In this way a test filter ^ is introduced and the expression is obtained

$$\tau_{SGS} = \rho\left(\widehat{\bar{u}_i \bar{u}_j} - \widehat{\bar{u}}_i \widehat{\bar{u}}_j\right). \tag{2.31}$$

A dynamic procedure can be implemented in order to calibrate the model parameter, by applying a test filter as explained for the similarity case. Considering the new filtered field τ_{SGS} can be explicitly computed using 2.29 and it can be compared with the model estimation for this filtered field. The parameter can then be determined and applied to the original resolved field, the underlying assumption being that the same parameter is valid for both the LES and the LES on the coarser numerical grid.

DNS

The Navier Stokes equations are solved without the consideration of any averaging, neither spatial nor temporal, so that all the turbulent structures are solved and the introduced error obeys only to the discretization. Considering homogeneous isotropic turbulence, the amount of points in each direction should at least be L_t/η, that scales as $Re_t^{3/4}$. Due to the high computational cost, DNS simulations are limited to low Re_t and simple geometrical configurations. Among the key advantages they offer is high fidelity data, which in many cases results either quite difficult when not impossible to measure experimentally, the possibility of isolated parameter study even under unphysical conditions, and finally access to instantaneous flow visualizations, which might give insight into the problems physics, as well as statistical data.

In DNS every time step should be accurately resolved, which sets the constraint of small time step and stable time advance, generally solved employing explicit methods. Hence Runga-Kutta methods are often used, while Crank-Nicholson is chosen for the case of implicit terms. Regarding the discretization, due to the large wavenumber range, the computation of the accuracy of the approximation based on the method order is no longer appropriate, and an error estimation based on the computation of the energy spectrum might be more convenient.

2.2 Numerical Methods

2.2.1 Finite volume method

The finite volume method is based on the integral form of the conservation equation as in Equation 2.1. The numerical grid divides the domain into control volumes CV. The grid points define the cell limits and not the computational node, which is generally located at the cell center, i.e. node centered, though other strategies are possible as well, e.g. face centered approach. The discretized equation is then obtained by approximating the surface and volume integrals.

The surface integral is the sum of the fluxes through each one of the faces:

$$\oint_S f \cdot dS = \sum_k \int_{S_k} f \cdot dS \,, \tag{2.32}$$

where the flux can be convective $f^{conv} = \rho \mathbf{u}\varphi \cdot \mathbf{n}$ or diffusive $f^{diff} = \Gamma \nabla \varphi \cdot \mathbf{n}$. It will be assumed that only the variable φ is unknown.

To compute the integral, approximations are introduced in two levels: first the flux value is computed considering one or more locations on the face, second the node values are used to estimate the face values. Concerning the first level, the midpoint rule is a second order accuracy approximation, where the flux is the product of the face center value and the face area. Thus for a given cell face i with center c

$$F_i = \int_{S_i} f \cdot dS \approx f_{i,c} \cdot S_i \,. \tag{2.33}$$

In order to compute the convective and diffusive flux terms on the cell face, the second level approximations based on the cell values are done. On the upwind interpolation or upwind differencing scheme (UDS) φ at the face i equals the value at the upstream node. This is equivalent to a backward or forward difference for the first derivative, thus a first order approximation. As a result this is unconditionally bounded, but numerically diffusive.

The linear interpolation or central differencing scheme (CDS) on the contrary considers the two nearest nodes for the estimation. It has second order accuracy, and consequently it might produce oscillatory solutions. Defining with P the computational cell, and I the adjacent cell sharing the face i, the approximation of φ_i employing CDS reads

$$\varphi_i = \varphi_I \lambda_i + \varphi_P (1 - \lambda_i) \,, \tag{2.34}$$

and the linear interpolation factor is

$$\lambda_i = \frac{x_i - x_P}{x_I - x_P} \,. \tag{2.35}$$

The same procedure can be applied to estimate the gradient with second order accuracy. Thus the diffusive flux reads

$$\left(\frac{\partial \varphi}{\partial x} \right)_i = \frac{\varphi_I - \varphi_P}{x_I - x_P} \,, \tag{2.36}$$

The volume integral can be approximated as the product of Q at the cell center and the cell volume.

$$\int_\Omega Q_V \, d\Omega \approx Q_c \Omega_c \,, \tag{2.37}$$

No interpolation is then needed, for Q constant or linearly varying within CV, the obtained value is exact; otherwise second order accuracy is obtained.

2.2.2 Pressure-velocity coupling

OpenFOAM offers three segregated solution strategies in order to handle the pressure-velocity coupling: SIMPLE [18], which has been conceived for steady-state problems, PISO [71] which is typically used for transient flows, and PIMPLE, a merged PISO-SIMPLE method.

Starting from the discretized momentum equation, it can be expressed as

$$A_P u_p = b_P - \sum_l A_N u_N - \left(\frac{\partial p_d}{\partial x_i} \right)_P = H(u) - \left(\frac{\partial p_d}{\partial x_i} \right)_P \,, \tag{2.38}$$

where P is an arbitrary computational cell, N the neighbor cells considered in the discretization, A_P the matrix diagonal coefficients and H contains the non-diagonal coefficients (mainly convective and diffusive) plus the source terms (including the explicit contributions of u and the body forces or other linearized terms). It is assumed that only the velocity is unknown, and the values of the previous time step are employed for all the other terms. In OpenFOAM all this information except p_d will be saved on the $UEqn$ matrix.

The first step consists on the velocity computation at $m*$ in the so-called momentum predictor

$$u_p^{m*} = \frac{b_P^{m-1} - \sum_l A_N u_N^{m*}}{A_P} - \frac{1}{A_P} \left(\frac{\partial p_d}{\partial x_i} \right)_P^{m-1} = \frac{H(u)^{m*}}{A_P} - \frac{1}{A_P} \left(\frac{\partial p_d}{\partial x_i} \right)_P^{m-1} \,. \tag{2.39}$$

Due to the sequential approach, the pressure at p^{m*} is unknown and thus the value at the previous time step p^{m-1} is employed. It follows that this velocity does not satisfy the continuity equation, theferore a correction is applied

$$u^m = u^{m*} + u' \, ,$$ (2.40)

$$p^m = p^{m-1} + p' \, .$$ (2.41)

The correction velocity can be expressed as

$$u'_p = \frac{-\sum_l A_N u'_N}{A_P} - \frac{1}{A_P}\left(\frac{\partial p_d}{\partial x_i}\right)'_P = \frac{H(u)'}{A_P} - \frac{1}{A_P}\left(\frac{\partial p_d}{\partial x_i}\right)'_P \, ,$$ (2.42)

and the final velocity

$$u^m_p = \frac{H(u)^{m*}}{A_P} + \frac{H(u)'}{A_P} - \frac{1}{A_P}\left(\frac{\partial p_d}{\partial x_i}\right)^m_P \, .$$ (2.43)

Expressing continuity equation in terms of u^m

$$\frac{\partial \rho}{\partial t} + \frac{\partial \rho u^m_i}{\partial x_i} = 0 \, ,$$ (2.44)

and then replacing 2.43 into 2.44 the expression is obtained

$$\frac{\partial \rho}{\partial t} + \frac{\partial}{\partial x_i}\left(\frac{\rho}{A_P}\left(H(u)^{m*} + H(u)'\right)\right) - \frac{\partial}{\partial x_i}\left(\frac{\rho}{A_P}\left(\frac{\partial p_d}{\partial x_i}\right)^m_P\right) = 0 \, .$$ (2.45)

The resolved $p_{d,P}$ can then be substituted in 2.43 and so u^m_P is updated. Different alternatives exist to handle the term $H(u')$. On the SIMPLE algorithm its contribution is neglected, i.e. $H(u') = 0$, while on the SIMPLEC algorithm an explicit relation is employed. An iterative procedure is employed where the update of the coefficients and source matrices, the velocity prediction, and the pressure correction intercalate until convergence is reached.

The PISO algorithm follows a different strategy as it employs one implicit momentum predictor, and more than one explicit corrector steps. In this case a modified H term is employed

$$H = H(u)^{m*} + r \, ,$$ (2.46)

where r comes from the time discretization. The pressure update equation becomes

$$\frac{\partial \rho}{\partial t} + \frac{\partial}{\partial x_i}\left(\frac{\rho H}{A_P}\right) - \frac{\partial}{\partial x_i}\left(\frac{\rho}{A_P}\left(\frac{\partial p_d}{\partial x_i}\right)^m_P\right) = 0 \, .$$ (2.47)

The velocity u^m can be then updated with p^m, and the correction procedure can be repeated, to resolve the coupling in one timestep. For the PIMPLE algorithm, inner PISO iterations are embedded into outer SIMPLE iterations, where the properties and matrix coefficients are updated, and the sequence is repeated until satisfying a given tolerance in each timestep.

2.3 Combustion Theory

This section presents fundamental concepts first for premixed and then for non-premixed combustion. Subsequently some turbulent combustion models are briefly explained. The presentation does not aim to comprehensively cover the state of the art, but the selected models are thought to be relevant to the present study due to their affinity with the non-premixed FTACLES formalism.

2.3.1 Premixed combustion

General concepts

Combustion can be defined as premixed when fuel and oxidizer are completely mixed before the reaction takes place. This can be carried out when the temperature lies below a given threshold, whose value depends on the composition, and for which the reactions can said to be frozen[110]. If the mixture is within the flammability limits, the addition of a heat source will start combustion. This generates a system with two clearly defined states, i.e. burnt and unburnt, separated by the propagating flame front. Considering a one-dimensional flame configuration, if transients are ignored from the continuity equation, the following relation can be obtained

$$\rho u = constant = \rho_0 s_l^0 \,, \tag{2.48}$$

where ρ_0 is the unburnt density and s_l^0 the laminar flame speed. This quantity corresponds to the velocity at which the flame front propagates normal to itself in the unburnt direction, it is a flame property, and depends on the composition, temperature and pressure. One definition for s_l^0 can be obtained by integrating the fuel conservation equation from $x = -\infty$ to $x = \infty$

$$s_l^0 = -\frac{1}{\rho_0 Y_F^0} \int_{-\infty}^{\infty} \dot{\omega}_F dn \,. \tag{2.49}$$

In this case the diffusion term vanishes at the extremes and s_l^0 reflects the velocity with which the flame burns the reactants.

Additional to a velocity scale, a premixed flame can be as well characterized in terms of a lenghtscale. Different laminar flame thickness definitions have been employed e.g. relating the thermal diffusivity and the laminar flame speed[26]

$$\delta_l^0 = \frac{\lambda}{\rho c_p s_l^0} \,, \tag{2.50}$$

or considering the temperature gradient on the flame front[135]

$$\delta_l^0 = \frac{T_2 - T_1}{max\left(\left|\frac{\partial T}{\partial x}\right|\right)} \,. \tag{2.51}$$

The structure on a premixed flame can be described by means of a reaction progress variable c. It depicts the reaction evolution from 0 in the unburnt to 1 in the burnt region and should fulfill that $\nabla c \neq 0$[142].

$$c = \frac{\varphi - \varphi_u}{\varphi_b - \varphi_u} \,. \tag{2.52}$$

where φ corresponds to c definition, which normally includes a combination of reactive scalars such as chemical species or temperature [68]. Numerous definitions have been proposed in literature, e.g. $\varphi = Y_{CO_2} + Y_{H_2O}$[114], $\varphi = Y_{CO_2} + Y_{H_2O} + Y_{CO} + Y_{H_2}$[65], $\varphi = Y_{CO_2} + Y_{CO}$[52], or $\varphi = \alpha_{CO_2}Y_{CO_2} + \alpha_{H2O}Y_{H_2O} + \alpha_{H_2}Y_{H_2}$[140], where α_i are weighting coefficients e.g. molecular weight MW_i. Assuming equal species diffusivities, an equation might be then written for c which will be then employed for combustion simulation.

The structure of a premixed flame corresponding to a $CH_4 - Air$ mixture at $Z = 0.056$ is shown on figure 2.2. The progress variable in this case has been defined as the algebraic sum of CO_2 and CO. Three normalized variables are plotted first in physical space and subsequently as function of c. The temperature and the main species mass fractions, e.g. CH_4, O_2 and CO_2, evolve monotonically along the physical domain, and so do thermophysical properties such as μ, ρ and λ. It is important to recall how the chemical activity is instead concentrated in a specific region, in which the source term reaches a maximum, and then decreases in both directions, burned and unburned products.

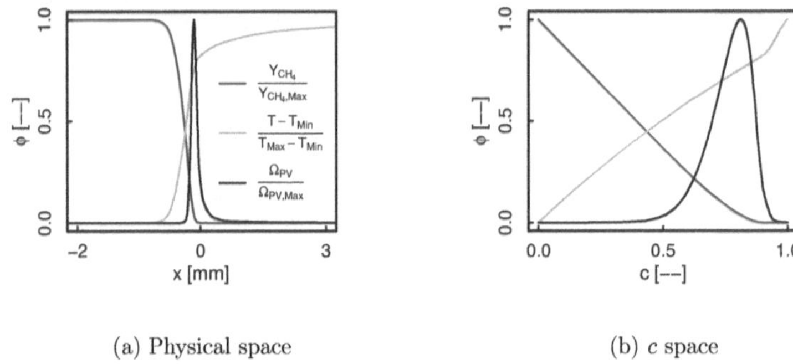

<table>
<tr><td>(a) Physical space</td><td>(b) c space</td></tr>
</table>

Figure 2.2: Premixed flame normalized CH_4, T and ω_{PV} profiles in physical and c space.

Turbulent combustion models

A problem that is faced when solving premixed flames on a typical LES mesh is that normally the flame thickness δ_l^0 lies in the range $[0.1 - 1]$mm[146], it is smaller than the filter size Δ_x, and consequently can not be resolved. The difficulty that arises from this condition is that since the chemical source term presents a highly non-linear behavior, its value can not be simply estimated as a function of the variable's mean value in the cell. Modeling strategies should be then developed, among which a well known approach consists to increase the flame thickness, so that its resolution on the numerical grid becomes possible. Two different models based on this principle are shown in this section: the artificial thickening of a flame and the direct filtering of a flame. Even though their formulation and justification differs, from the physical point of view both of the approaches intend to increase the flame front thickness and deal with the consequent modification of the source term. They are presented in this section since useful concepts employed in these models will then be as well applied for the case of the non-premixed FTACLES model.

Thickened Flame Model

One strategy to resolve a premixed flame front in a coarser mesh consists to increase the flame thickness by a factor F while conserving the laminar flame speed unaltered [17],[107]. This idea has been developed for LES [29] and the modified turbulence interaction has been addressed employing linear and power law wrinkling models in dynamic [23] and non-dynamic formulations[22]. The formalism has been coupled with FGM [64] [77] [123], detailed kinetics [49], it has been succesfully employed on stratified combustion [121] [137] and spray flames [125] [43]. The model exploits the following properties which hold for a premixed flame

$$s_l^0 \propto \sqrt{D\bar{\omega}}, \tag{2.53}$$

$$\delta_l^0 \propto \frac{D}{s_l^0}, \tag{2.54}$$

where D is the molecular diffusivity and $\bar{\omega}$ the mean reaction rate. Multiplying D and at the same time dividing $\bar{\omega}$ by a factor F maintains the same s_l^0, with the benefit that if F is chosen sufficiently large (from 5 to 30) [29], the flame can then be resolved on the LES mesh. In this way equation 2.7 becomes

$$\frac{\partial \bar{\rho}\tilde{c}}{\partial t} + \frac{\partial}{\partial x_i}(\bar{\rho}\tilde{u}_i\tilde{c}) = \frac{\partial}{\partial x_i}\left(F\bar{\rho}\tilde{D}\frac{\partial \tilde{c}}{\partial x_i}\right) + \frac{\overline{\dot{\omega}}_c}{F}. \tag{2.55}$$

One of the challenges that however emerges from this approach is the modification of the turbulence-flame interaction. The eddies smaller than $F\delta_l^0$ do not interact with the flame anymore, while the behavior of those bigger than the threshold is changed. This causes the flame to be less sensitive to turbulence motions, at the same time that the sensitivity to strain is increased.

Departing from the flame surface density balance equation[145][138], filtering in an LES context, and assuming equilibrium at the SGS level between flame surface and turbulence, the wrinkling factor Ξ can be defined. It describes the subgrid scale flame surface divided by the projection of the flame surface in the propagating direction as

$$\Xi \approx 1 + \alpha\frac{\Delta_e}{s_l^0}a_{T,s}, \tag{2.56}$$

where α is a model constant to be determined and the term $a_{T,s}$ corresponds to the SGS strain rate, which needs to be estimated as well. One possibility is to estimate $a_{T,s}$ as the ratio of the SGS turbulent velocity u'_{Δ_e} and the filter size Δ_e, as done in the G-equation approach [70]

$$\Xi \approx 1 + \alpha\frac{\Delta_e}{s_l^0}\frac{u'_{\Delta_e}}{\Delta_e} = 1 + \alpha\frac{u'_{\Delta_e}}{s_l^0}. \tag{2.57}$$

The linear model proposed in[29] estimates the wrinkling of the SGS flame front as

$$\Xi = 1 + \alpha\Gamma\left(\frac{\Delta_e}{\delta_l^1}, \frac{u'_{\Delta_e}}{s_l^0}\right)\frac{u'_{\Delta_e}}{s_l^0}, \tag{2.58}$$

$$\Gamma\left(\frac{\Delta_e}{\delta_l^1}, \frac{u'_{\Delta_e}}{s_l^0}\right) = \frac{a_{T,s}}{u'_{\Delta_e}/\Delta_e} = 0.75 exp\left[-\frac{1.2}{(u'_{\Delta_e}/s_l^0)^{0.3}}\right]\left(\frac{\Delta_e}{\delta_l^1}\right)^{2/3}, \tag{2.59}$$

$$\alpha = \beta\frac{2ln(2)}{3c_{ms}\left[Re_t^{1/2} - 1\right]}, \tag{2.60}$$

being β a model constant of the order of unity, $c_{ms} = 0.28$, and Re_t the turbulent Reynolds number. The subgrid scale turbulent velocity is computed employing a similarity assumption

$$u' = c(\bar{u} - \tilde{\bar{u}}) = OP(\bar{u}), \tag{2.61}$$

where $\tilde{\bar{u}}$ corresponds to a local average of the resolved velocity field $\bar{u}$. The operator considers the rotational part of the velocity field, and so the influence of the dilatational component is omitted. This resulting into

$$u'_{\Delta_e} = |u'| = c_2\Delta_x^3|\nabla^2(\nabla \times \bar{u})|, \tag{2.62}$$

where the constant c_2 is mainly devoted to correct the estimation between Δ_e and Δ_x, and for which from an energy spectrum analysis for a wide number of turbulent Reynolds numbers, a value of 2 was assigned.

Since in a thickened flame the actual wrinkling of the flame front is underestimated, the efficiency factor ξ considering the wrinkling both of the original as of the thickened flame is defined as

$$\xi = \frac{\Xi(\delta_l^0)}{\Xi(\delta_l^1)} = \frac{1 + \alpha\Gamma\left(\frac{\Delta_e}{\delta_l^0}, \frac{u'_{\Delta_e}}{s_l^0}\right)\frac{u'_{\Delta_e}}{s_l^1}}{1 + \alpha\Gamma\left(\frac{\Delta_e}{\delta_l^1}, \frac{u'_{\Delta_e}}{s_l^0}\right)\frac{u'_{\Delta_e}}{s_l^0}}. \tag{2.63}$$

In this way, the underestimation of the flame wrinkling is compensated by an increase in the flame speed through ξ. The modified flame corresponds to one with a thickness of $\delta_l^1 = F\delta_l^0$ and a speed of $s_l^1 = \xi s_l^0$.

A non negligible drawback of this model in its original version is the modification of the diffusion coefficient all along the domain, i.e. even far away from the flame front[44]. Since the species concentration might change not only due chemical reactions but as well due to pure mixing, the modification of the diffusion term without its source term counterpart, as in these cases might produce an unphysical behavior. One alternative that has then been developed in order to solve this problem is the use of a sensor that detects the presence of the flame and consequently limits the action of the model to the region close to the flame front. Durand and Polifke [44] propose then a definition based on the progress variable as

$$S(PV) = 16\left[PV\left(1 - PV\right)^2\right]. \tag{2.64}$$

Leading to the final expression for the species transport equation in the thickened flame context

$$\frac{\partial\bar{\rho}\tilde{c}}{\partial t} + \frac{\partial}{\partial x_i}(\bar{\rho}\tilde{u}_i\tilde{c}) = S\frac{\partial}{\partial x_i}\left(\xi F\bar{\rho}\tilde{D}\frac{\partial\tilde{c}}{\partial x_i}\right) + (1 - S)\frac{\partial}{\partial x_i}\left(\bar{\rho}(\tilde{D} + \tilde{D}_t)\frac{\partial\tilde{c}}{\partial x_i}\right) + \frac{\xi}{F}\bar{\dot{\omega}}_c. \tag{2.65}$$

FTACLES

The Filtered Tabulated Chemistry for LES[53] is a strategy in which directly filtering one-dimensional flamelets and carefully addressing the closure terms, both the flame propagation speed as well as the filtered flame structure can be correctly retrieved on an LES approach. The model has been combined with dynamic wrinkling procedures [130], and employed on stratified flames with[97][98] and without adiabatic effects[6] as well as on ignition process assessment[112].

The formalism is based on the fact that if there is no wrinkling at the subgrid scale, the propagation speed of the filtered flame front s_Δ does not differ from the laminar flame speed s_l^0. This statement can be expressed as

$$\rho_0 s_\Delta = \int_{-\infty}^{\infty} \bar{\rho}\tilde{\dot{\omega}}_c(x)dx = \int_{-\infty}^{\infty} \rho\dot{\omega}_c(x)dx = \rho_0 s_l^0 . \tag{2.66}$$

This feature is not fulfilled if $\tilde{\dot{\omega}}_c$ is approximated employing a presumed β-PDF and the filter size, used for the estimation of $\tilde{c}$ and $\widetilde{cc}$ on PDF construction, is larger than the flame front. For this reason the model estimates $\tilde{\dot{\omega}}_c$ through direct filtering as:

$$\tilde{\dot{\omega}}_c = \frac{1}{\bar{\rho}} \int_{-\infty}^{\infty} \rho(x')\dot{\omega}_c(x')F(x - x')dx' . \tag{2.67}$$

Due to the properties of the filter function $F(x)$, i.e. $\int_{-\infty}^{\infty} F(x)dx = 1$, equation 2.66 holds and in this way the adequate values for s_l^0 are obtained independent of the filter size employed.

The filtering operation is then applied to the entire flamelet solution. This permits to express the transport equation 2.7 in terms of the progress variable as

$$\frac{\partial \bar{\rho}\tilde{c}}{\partial t} + \frac{\partial}{\partial x_i}(\bar{\rho}\tilde{u}_i\tilde{c}) = \frac{\partial}{\partial x_i}\left(\alpha_c\bar{\rho}\tilde{D}\frac{\partial \tilde{c}}{\partial x_i}\right) + \bar{\rho}\tilde{\dot{\omega}}_c + \dot{\omega}_{conv} , \tag{2.68}$$

where $\dot{\omega}_{conv}$ is the convective correction and α_c the diffusion correction. These terms, unclosed in a filtered approach like LES can therefore be explicitly computed as:

$$\dot{\omega}_{conv} = -\frac{\partial}{\partial x_i}(\overline{\rho u_i c} - \bar{\rho}\tilde{u}_i\tilde{c}) = -\rho_0 s_l^0 \left(\frac{\partial \overline{c^*}}{\partial x^*} - \frac{\partial \tilde{c}^*}{\partial x^*}\right) , \tag{2.69}$$

$$\alpha_c = \left(\overline{\rho D\frac{\partial c^*}{\partial x^*}}\right) \Big/ \left(\bar{\rho}\tilde{D}\frac{\partial \tilde{c}^*}{\partial x^*}\right) , \tag{2.70}$$

where * corresponds to quantities issued from the one-dimensional flamelets.

In order to take into account the subgrid scale wrinkling different approaches have been used that range from analytical models as that of Colin [29] up to much more complex dynamic models as in [130].

The consideration of ξ and S finally gives

$$\begin{aligned}
\frac{\partial \bar{\rho}\tilde{c}}{\partial t} + \frac{\partial}{\partial x_i}(\bar{\rho}\tilde{u}_i\tilde{c}) &= S\frac{\partial}{\partial x_i}\left(\xi\alpha_c\bar{\rho}\tilde{D}\frac{\partial \tilde{c}}{\partial x_i}\right) + \xi\bar{\rho}\tilde{\dot{\omega}}_c \\
&\quad +(1 - S)\frac{\partial}{\partial x_i}\left(\rho(\tilde{D} + \tilde{D}_t)\frac{\partial \tilde{c}}{\partial x_i}\right) + \xi\dot{\omega}_{conv} .
\end{aligned} \tag{2.71}$$

2.3.2 Non-premixed combustion

Non-premixed combustion can be defined as the process in which fuel and oxidizer, being injected in separate streams, mix through convective and diffusion phenomena and then burn. Another way to call the flames generated in these conditions is diffusion flames, which comes from the fact that the chemical time scale is much smaller than the time required for the diffusive and convective transport, i.e. the mixing, so that diffusion becomes the rate controlling process [111]. As a matter of fact in general both convective and diffusive time scales are of the same order of magnitude [110], and this feature, namely that non-premixed combustion is dominated by transport phenomena, can be considered as a first hint to the idea of parameterizing the process evolution in terms of the mixing between the composing streams.

Mixture Fraction

One way to express the mixture fraction concept is based on chemical elements as they are conserved in a reaction process, contrary to species. The mass fraction of an element j is

$$Z_j = \frac{m_j}{m} = \sum_{i=1}^{n} \frac{a_{ij} MW_j}{MW_i} Y_i \,, \tag{2.72}$$

where m_j is the total mass of the element j, m the mass in the system , MW the molecular weight and a_{ij} the number of j atoms in species i with mass fraction Y_i.

Considering binary diffusion and that all diffusivities are equal, $D_i = D$, the balance equation for the mass fraction of an element j can be expressed as

$$\frac{\partial \rho Z_j}{\partial t} + \frac{\partial \rho u Z_j}{\partial x} = \frac{\partial}{\partial x}\left(\rho D \frac{\partial Z}{\partial x}\right). \tag{2.73}$$

The equation describes the evolution of a passive scalar (it changes due to diffusion and convection but not due to reaction).

Given a complete combustion reaction

$$\nu_F C_{n_C} H_{n_H} + \nu_{O_2} O_2 \rightarrow \nu_{CO_2} CO_2 + \nu_{H_2O} H_2O \,, \tag{2.74}$$

where n_C and n_H are respectively the number of carbon and hydrogen atoms in the fuel, and assuming for notation simplicity $\nu_F = 1$, it holds that

$$\frac{Z_C}{n_C W_C} = \frac{Z_H}{n_H W_H} = \frac{Y_{F,u}}{W_F}, \quad Z_O = Y_{O_2,u}\,. \tag{2.75}$$

The coupling function β, which vanishes at stoichiometric conditions can then be introduced as

$$\beta = \frac{Z_C}{n_C W_C} + \frac{Z_H}{n_H W_H} - 2\frac{Z_O}{\nu_{O_2} W_{O_2}}\,, \tag{2.76}$$

and the definition for conserved scalar as in[15] is obtained. This can be then normalized between 0 and 1 as then defined in [12]

$$Z = \frac{\beta - \beta_2}{\beta_1 - \beta_2}\,. \tag{2.77}$$

The mixture fraction concept permits to reduce the number of transported variables, by assuming that all species and temperature can be described as

$$Y_i = Y_i(Z,t) \quad \text{and} \quad T = T(Z,t). \tag{2.78}$$

This permits to decouple diffusion flame computations into two problems: mixing and flame structure problem[120]. The mixture fraction field should be resolved as a function of spatial coordinates and time $Z(x,t)$, while the links between flame variables (mass fractions $Y_i(Z)$ and temperature $T(Z)$) are used to construct all flame variables.

Laminar diffusion flamelet

Assuming high enough Z gradients, combustion takes place in a thin layer around the stoichiometric surface $Z(x,t) = Z_{st}$. The consideration of the former plus the surrounding inert mixing region defines a laminar diffusion flamelet[84],[110]. The flamelet concept proposes that the behavior of a real laminar flame can be understood as the combination of a set of laminar flamelets, if these last ones span the whole range in which the original flame moves.

A coordinate transformation of the Crocco-type with Z, locally normal to the stoichiometric surface, replacing the x_1 coordinate, $Z_2 = x_2$, $Z_3 = x_3$ and $\tau = t$ is applied to the species conservation equation (2.7). Considering that the Z_2 and Z_3 related derivative terms are of lower order with respect to those linked to Z, and further assuming equal diffusivity for all the species, the diffusion flamelet formalism results into a one-dimensional flame description. Thus the species conservation reads

$$\frac{1}{2}\rho\chi\frac{\partial^2 Y_i}{\partial Z^2} = -\omega_{Y_i}, \tag{2.79}$$

where the scalar dissipation rate $\chi = f(Z,t)$ reflects the influence of the flow field and is defined as

$$\chi = 2D\frac{\partial Z}{\partial x}\frac{\partial Z}{\partial x}. \tag{2.80}$$

Equation (2.79) sets a starting point from which the species mass fractions and the temperature can be related to the mixture fraction. If a unique functional dependence of χ on Z is assumed, then the expression can be fully defined and the following relation can be written

$$\varphi = \varphi(Z,\chi). \tag{2.81}$$

The scalar dissipation rate varies as $\chi = f(Z,t)$ according to the expression

$$\chi = \chi_0 F(Z), \tag{2.82}$$

where χ_0 corresponds to the local peak of χ within the layer and is assumed to be independent from Z, and F is then given by

$$F(Z) = exp(-2[erf^{-1}(2Z-1)]^2). \tag{2.83}$$

When solving equation (2.81) for different strain rates, it results useful to characterize each flamelet in terms of the controlling parameters at the stoichiometric condition rather than at χ_0 position. The commonly used procedure then is to specify the desired χ_{st}, scale it by the reference value at $F(Z_{st})$, whose shape is shown in Figure 2.3, and in this way determine the corresponding χ_0.

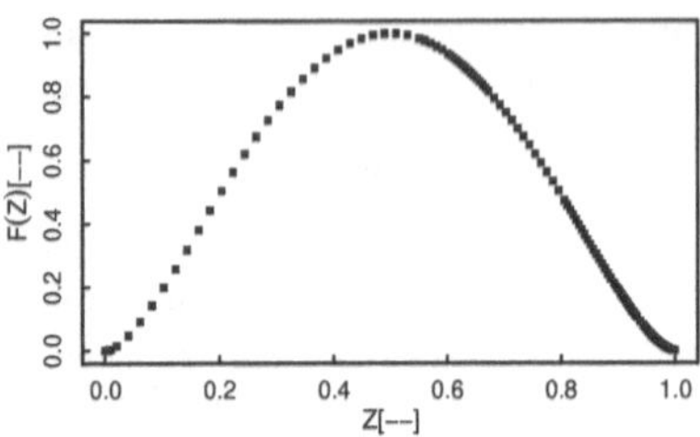

Figure 2.3: $F(Z)$ profile

Turbulent combustion models

The flamelet model has been applied for turbulent combustion both in RANS[91][103], as in LES[102][93] context. A well known approach consists to use presumed PDF to estimate the turbulence-chemistry interaction, though other strategies, e.g. conditional source term estimation[16] or artificial inteligence[48] have been explored as well. Within the presumed PDF framework, the approach has been further developed considering among others the number of parameterizing variables[154][151], the addition of unsteady flamelets[116][115], and the type of presumed functions[27]. Two models are therefore exemplarily presented, as they offer a representative view of the varying complexity in the solution computation.

Steady flamelet model

The laminar flamelet model for LES as proposed in[30] takes into account the turbulence effect through a presumed PDF approach, therefore estimating the filtered quantities as

$$\tilde{\varphi} = \int_0^1 \int_{\chi^{min}}^{\chi^{max}} \varphi(Z,\chi) P(Z,\chi) d\chi dZ , \qquad (2.84)$$

where $P(Z,\chi)$ is the joint PDF. Due to the independence between χ_0 and Z, χ is substituted for χ_0. Then further considering φ as a weak, approximately linear function of χ_0, its value can be approximated using the first two terms of a Taylor expansion about the average of χ_0. In this way equation (2.84) becomes

$$\tilde{\varphi} = \int_0^1 \varphi(Z,\tilde{\chi}_0) P(Z) dZ , \qquad (2.85)$$

where χ_0 is estimated by averaging equation (2.82) as

$$\tilde{\chi}_0 = \frac{\tilde{\chi}}{\int_0^1 F(Z) P(Z) dZ} . \qquad (2.86)$$

The PDF for Z is modeled employing a β distribution, as it can quite adequately describe passive scalars[58]

$$P(Z) = \frac{Z^{a-1}(1-Z)^{b-1}}{B(a,b)} , \qquad (2.87)$$

$$a = \tilde{Z}\left(\frac{\tilde{Z}(1-\tilde{Z})}{Z_v^2} - 1\right) \geq 0 , \qquad (2.88)$$

$$b = a\left(\frac{1}{\tilde{Z}} - 1\right) \geq 0. \tag{2.89}$$

Being $B(a,b)$ the beta function, and Z_v^2 is the variance. The filtered values are precomputed and tabulated considering the the full range of $\tilde{Z}$, Z_v^2 and $\tilde{\chi}$ expected from the LES. The modeling strategy therefore additional to the filtered mixed fraction $\tilde{Z}$, considers its subgrid scale variance Z_v^2 and the filtered scalar dissipation rate $\tilde{\chi}$ which can be either transported or algebraically estimated[33].

Flamelet/Progress Variable

When solving equation (2.79) and plotting the stoichiometric temperature T_{st} vrs the stoichiometric scalar dissipation rate χ_{st} the so called S-shaped curve is obtained. The presence of three branches can then be recognized: the lower one represents weakly reactive states, the middle one unstable solutions and the upper one vigorously burning states[105]. The laminar steady flamelet method solves the problem of having multiple solutions for certain χ values by considering only the burning conditions for $\chi < \chi_q$ and moving to the lower branch just after the extinction. This implies that through this unique parametrization in terms of χ not the whole solution space can be represented, making it not possible to describe the intermediate states between fully burning and fully extinguished, as in re-ignition phenomena. The FPV approach[113][114] proposes to overcome this limitation by adding a reaction progress parameter λ, based on a reaction variable c, to the passive scalar Z. The solution to equation (2.79) can be now expressed as

$$\varphi = \varphi(Z, \lambda). \tag{2.90}$$

The main difference respect to equation (2.81) is that all solutions of the steady state flamelet equations are included, since this parametrization permits to uniquely identify each single flame state along the S-shaped curve.

Different choices can be made for the variable c, e.g. a linear combination of major reaction products or temperature; while for λ a common practice is to choose the value at the stoichiometric condition. This flamelet parameter λ allows to uniquely identify each flamelet in the library since it describes any (Z, c) combination in a given flamelet, implying then as well that λ is independent from Z.

In order to obtain any filtered combustion variable $\tilde{\varphi}$ such as the chemical species, temperature, density, or the source term of the progress variable, the chemical state relationships should be now integrated over the joint PDF of Z and λ

$$\tilde{\varphi} = \int \varphi(Z, \lambda)\tilde{P}(Z, \lambda)dZd\lambda. \tag{2.91}$$

By using Bayes' theorem this joint PDF can be written in terms of a conditional and a marginal PDF as

$$\tilde{P}(Z, \lambda) = \tilde{P}(\lambda|Z)\tilde{P}(Z), \tag{2.92}$$

but due to the statistical independence between Z and λ, the relation can be expressed as

$$\tilde{P}(Z, \lambda) = \tilde{P}(Z)P(\lambda). \tag{2.93}$$

As has already been mentioned above, a β distribution is considered appropriate for $\tilde{P}(Z)$. Then the marginal PDF of λ is described using a delta function.

Even though the definition of λ simplifies the presumed PDF closure model, the solution of its transport equation requires the modeling of various unclosed terms, which is non trivial [67]. For this reason thermochemical species are finally expressed the following state relation

$$\varphi = \varphi(Z, c)\,. \tag{2.94}$$

The FPV model has been extended and modified in order to improve its performance in specific conditions, e.g. by employing a more complex PDF closure through a statistically most likely distribution (SMLD) for λ[67]. Furthermore, to better account for unsteady phenomena both λ and χ_{st} have been considered as tabulating parameters[117], so that

$$\varphi = \varphi(Z, \lambda, \chi_{st})\,. \tag{2.95}$$

Both Z and λ were already proved to be independent, and the behavior of χ_{st} respect to Z had as well been shown to be independent in equation (2.82). In this way the joint PDF can be finally expressed as

$$\tilde{P}(Z, \lambda, \chi_{st}) = \beta(Z; \tilde{Z}, \widetilde{Z''^2})\delta(\lambda - \lambda^*)\delta(\chi_{st} - \chi_{st}^*)\,, \tag{2.96}$$

where for computational efficiency, once built, the table is reinterpolated replacing λ by $\tilde{c}$ and χ_{st} for $\tilde{\chi}$ which are all known parameters of the LES simulation.

Chapter 3

Non-premixed FTACLES

The FTACLES model was originally developed for premixed combustion, and it employs flamelet direct filtering to deal with the chemical source term closure problem on LES [53]. Given the flexibility and simultaneous simplicity due to the absence of additional assumptions that the concept offers, the model extension for turbulent diffusion flames without SGS wrinkling has been proposed by Coussement [32]. Premixed flames can be described in terms of characteristic velocity and length scales. The premixed FTACLES formalism is theoretically justified by the conservation of the laminar flame speed through the filtering operation. However, it is not straightforward to transpose this argument to diffusive flames, due to the lack of a characteristic velocity. It follows that the non-premixed FTACLES formalism is substantiated from a numerical point of view.

This chapter addresses two main questions: what does it mean to filter a non-premixed flamelet? How can this numerical operation be interpreted from a physical perspective? In order respond to this, first the filtering of a counterflow flamelet is explained from the theoretical point of view. The flamelet equations are introduced, the filtering operator is applied, and the derivation of the non-premixed FTACLES model is presented. The model is appraised considering an individual flamelet and the entire manifold. The undergoing transformations are identified and extensively discussed. Subsequently,a comparison is performed between the non-premixed FTACLES formalism and two turbulent combustion modeling strategies, namely the premixed FTACLES and the presumed PDF approach. Analogies are established and the specificities of the formalism are highlighted as well. The chapter distribution is as follows

- The equations describing the FTACLES model are first presented in Section 3.1.

- A characterization of non-premixed flames filtering is presented considering one individual flamelet and the whole manifold in Section 3.2.

- The non-premixed FTACLES model is compared against the premixed FTACLES in Section 3.3 and against the presumed β-PDF integration in Section 3.4.

- The model implementation is verified employing one and two-dimensional geometries in Section 3.5.

- Section 3.6 summarizes the main findings of the chapter.

3.1 Model description

This section introduces the model from the numerical point of view. The filtered transport equations for combustion modeling employing tabulated chemistry are presented. The closure of the parameterizing variables unresolved terms is done by filtering the flamelet equations. The parameterizing variables transport equation is extended first considering a flame sensor to verify the counterflow flame hypothesis, and then taking into account the SGS wrinkling.

In the FGM context, the flamelet equations[39] are one-dimensional expressions which describe mass, species and energy conservation in a flame adapted coordinate system

$$\frac{\partial \rho u}{\partial s} + \rho K = 0 \,, \tag{3.1}$$

$$\frac{\partial \rho u Y_i}{\partial s} + \rho K Y_i = \frac{1}{Le_i} \frac{\partial}{\partial s}\left(\frac{\lambda}{c_p}\frac{\partial Y_i}{\partial s}\right) + \dot{\omega}_{Y_i} \,, \tag{3.2}$$

$$\frac{\partial \rho u h}{\partial s} + \rho K h = \frac{\partial}{\partial s}\left[\frac{\lambda}{c_p}\frac{\partial h}{\partial s} + \sum_{1=1}^{N_s} h_i \frac{\lambda}{c_p}\left(\frac{1}{Le_i} - 1\right)\frac{\partial Y_i}{\partial s}\right] \,, \tag{3.3}$$

where s is the spatial coordinate perpendicular to the flame front, K the flame stretch rate, Y_i the mass fraction of species i, and h the enthalpy. Le_i is the Lewis number of species i, λ the thermal conductivity, c_p the specific heat at constant pressure, $\dot{\omega}_{Y_i}$ the chemical production rate, and N_s the total number of species.

The counterflow flame configuration can be used to describe diffusion flames, where the two dimensional flow field effects are taken into account through the local stretch rate K[38]. Thus, an equation for the transported stretch field is added [122]

$$\frac{\partial \rho u K}{\partial x} = \frac{\partial}{\partial x}\left(\frac{\mu \partial K}{\partial x}\right) + -2\rho K^2 + \rho_2 a^2 \,, \tag{3.4}$$

where a is the applied strain rate s^{-1} and ρ_2 the density, both at the oxidizer stream.

The solution at varying K generates a non-premixed low-dimensional manifold. For an adiabatic condition the thermodynamical and chemical properties can be uniquely parameterized as $\varphi = f(Z, c)$. This manifold is coupled to the flow field by solving the parameterizing variables transport equations together with the Navier-Stokes equations. Filtering the equations, as it would be in the LES context, the resulting system reads

$$\frac{\partial \bar{\rho}}{\partial t} + \frac{\partial}{\partial x_i}(\bar{\rho}\tilde{u}_i) = 0 \tag{3.5}$$

$$\frac{\partial \bar{\rho}\tilde{u}_j}{\partial t} + \frac{\partial}{\partial x_i}(\bar{\rho}\tilde{u}_i\tilde{u}_j) = -\frac{\partial}{\partial x_j}\bar{P} + \frac{\partial}{\partial x_i}\bar{\tau} - \frac{\partial}{\partial x_i}\bar{\tau}^t \tag{3.6}$$

$$\frac{\partial \bar{\rho}\tilde{\varphi}}{\partial t} + \frac{\partial}{\partial x_i}(\bar{\rho}\tilde{u}_i\tilde{\varphi}) = \frac{\partial}{\partial x_i}\left(\bar{\rho}\tilde{D}\frac{\partial \tilde{\varphi}}{\partial x_i}\right) + \bar{\dot{\omega}}_\varphi + \Omega_\varphi + \alpha_\varphi \,, \tag{3.7}$$

where φ corresponds to the parametrizing variables, D is the molecular diffusivity, $\bar{\dot{\omega}}_\varphi$ the unresolved chemical production term, and Ω_φ and α_φ are the unresolved convective and diffusive terms respectively.

Acknowledging that both Z and c result from linear combination of species, equation 3.7 can be rewritten in terms of the composing species, following equation 3.2. The LES filtering operation is approximated employing a Gaussian filter

$$F(x) = \frac{6}{\pi\Delta^2}^{1/2} exp\left(-\frac{6x^2}{\Delta^2}\right) . \tag{3.8}$$

This filter has been selected instead of for instance a more intuitive choice like the box filter, due to the filter and transfer functions analogous behavior in physical and in spectral space respectively, shown in Table 3.1. This results a quite convenient property when performing spectral analysis, e.g. energy cascade computations, contrary to the sinusoidal behavior of the box filter.

Table 3.1: Different filter types characterizing functions. κ the wavenumber and κ_c the cutoff wavenumber.

Filter	Filter function	Transfer function
Box	$\frac{1}{\Delta}H\left(\frac{1}{2}\Delta - \lvert x\rvert\right)$	$sin(\frac{1}{2}\kappa\Delta)/(\frac{1}{2}\kappa\Delta)$
Gaussian	$\frac{6}{\pi\Delta^2}^{1/2}exp\left(-\frac{6x^2}{\Delta^2}\right)$	$exp\left(-\frac{\kappa^2\Delta^2}{24}\right)$
Sharp spectral	$sin(\pi x/\Delta)/(\pi x)$	$H(\kappa_c - \lvert\kappa\rvert)$

Subsequently, each one of the budget terms in the transport equation can be exactly computed, e.g. source term, convective and diffusive fluxes, both for the unfiltered and for the filtered condition. Taking advantage of this feature, and assuming that there is no SGS wrinkling, Coussement[32] has proposed the non-premixed FTACLES model, that explicitly computes the unclosed terms on equation 3.7.

Referring to the flamelet solutions with the symbol *, the convective component is decomposed in terms of the contributions in flame front local coordinates (x_n, x_t) as:

$$\Omega_\varphi = \frac{\partial}{\partial x_n}\left(\overline{\rho^* u^* \varphi^*} - \bar{\rho}^* \tilde{u}^* \tilde{\varphi}^*\right) + \left(\overline{\rho^* K^* \varphi^*} - \bar{\rho}^* \tilde{K} \tilde{\varphi}^*\right) , \tag{3.9}$$

where u and K are the variables defined in the flamelet equations. Considering the vector normal to the flame front $n = \nabla\tilde{Z}/\lvert\tilde{Z}\rvert$, and assuming that ∇Z and $\nabla\varphi$ are aligned, the diffusive term becomes:

$$\alpha_\varphi = \frac{\partial}{\partial x_n}\left[\overline{\rho D^* \frac{\partial\varphi^*}{\partial x_n} sign\left(\frac{\partial Z^*}{\partial x_n}\right) n}\right] - \frac{\partial}{\partial x_n}\left[\bar{\rho}\tilde{D}^* \frac{\partial\tilde{\varphi}^*}{\partial x_n} sign\left(\frac{\partial\tilde{Z}^*}{\partial x_n}\right)\right] \tag{3.10}$$

Since on real configurations the counterflow flame hypothesis does not necessarily hold, a sensor S has been introduced to ensure this condition. This issue is carefully addressed in Section 4.2, where based on the underlying physics, an expression for S is proposed in equation 4.1. The sensor verifies the one-dimensionality of the flame structure, and is actually independent of the filtering process, resulting in the expression

$$\frac{\partial\bar{\rho}\tilde{\varphi}}{\partial t} + \frac{\partial}{\partial x_i}(\bar{\rho}\tilde{u}_i\tilde{\varphi}) = \frac{\partial}{\partial x_i}\left(\bar{D}\frac{\partial\tilde{\varphi}}{\partial x_i}\right) + \overline{\tilde{\omega}}_\varphi + S\left(\Omega_\varphi + \alpha_\varphi\right) . \tag{3.11}$$

A highly exploited approach in the context of turbulent premixed flamelet-based methods in order to overcome the difficulty of resolving the flame front on practical application computational meshes is to increase the flame thickness [17],[53]. Since the thickening operation decreases the efficiency of the vortices to augment the available flame surface, modeling strategies to consider the lost wrinkling over this modified flame have been developed and have proved to deliver satisfactory results [29] [22] [130].

Non-premixed flames present an analogous behavior: turbulence increases the flame surface area, scalar gradients, and consequently augments the rate of diffusive mixing [85]. Given that the non-premixed FTACLES method can be described as a flame thickening process, a flame wrinkling modeling strategy is considered. Nonetheless, due to the novelty of the problem, to the authors knowledge no expressions are available to take into account the modified turbulence-chemistry interaction on a thickened diffusion flame. To derive an efficiency function ξ for non-premixed flames at this step is highly needed. Since this task requires targeted DNS data that are out of the scope of this study, an expression based on the formulation proposed in [29] is rather adapted. It yields

$$\xi = \frac{\Xi(\delta_l^0)}{\Xi(\delta_l^1)} = \frac{1 + c_\xi \Gamma\left(\frac{\Delta_e}{\delta_l^0}, u'_{\Delta_e}\right) u'_{\Delta_e}}{1 + c_\xi \Gamma\left(\frac{\Delta_e}{\delta_l^1}, u'_{\Delta_e}\right) u'_{\Delta_e}}, \tag{3.12}$$

$$c_\xi = c_\alpha \cdot \alpha, \tag{3.13}$$

with following terminologies, Ξ is the wrinkling factor, δ_l^0 and δ_l^1 are the laminar flame thickness and the modified flame thickness, respectively, Δ_e the filter size and u'_{Δ_e} the subgrid scale turbulent velocity. c_ξ is a model constant with α defined as Equation 2.60 and c_α a factor related to the flame composition, $c_\alpha = 0.279$ for the current case. Γ estimates the ratio between the global effective strain rate $\langle a_T \rangle_s$ and u'_{Δ_e}/Δ_e. Note that equation 2.59 has been employed with $s_l^0|Z_{st}$ substituting s_l^0. The formulation is analogous to the approach based on r/δ_l^0 (r is the vortex size) followed in [96] when analyzing the vortex pair interaction with a laminar flame. The turbulent case equation reads

$$\frac{\partial \bar{\rho}\tilde{\varphi}}{\partial t} + \frac{\partial}{\partial x_i}(\bar{\rho}\tilde{u}_i\tilde{\varphi}) = \frac{\partial}{\partial x_i}\left(\xi\bar{D}\frac{\partial\tilde{\varphi}}{\partial x_i}\right) + S\left(\Omega_\varphi + \alpha_\varphi\right)$$
$$+ (1 - S)\frac{\partial}{\partial x_i}\left(D_{sgs}\frac{\partial\tilde{\varphi}}{\partial x_i}\right) + \xi\bar{\dot{\omega}}_\varphi. \tag{3.14}$$

The continuous activation of ξ, independent from S, highlights that the read variables are always filtered. Hence the flamelet structure is changed, or thickened, and so is the turbulence-chemistry interaction. This differs e.g. from the ATF model, where the selection between the thickened and the unmodified diffusivity is a function of S. On the contrary, two possibilities for the estimation of the turbulent flux emerge based on the sensor. The second term on the RHS of equation 3.14 describes the first scenario, where the sensor is active and the term is explicitly computed from the preprocessed flamelets. For the case where the counterflow hypothesis does not hold, the turbulent flux is approximated employing the subgrid scale diffusivity with a gradient diffusion hypothesis, as expressed by the third RHS term.

In order to adequately represent the underlying physics, the thickness should be defined in such a way that it increases with the filter size. For the case of strained non-premixed flamelets a commonly used way to estimate the flame thickness is based on conditional values at the stoichiometric condition, as $\delta_{\nabla Z_{st}} = 1/|\nabla Z_{st}|$. The inconvenience of this definition lies on the fact that the filtering operation extends the profile around the maximum value, which does not necessarily correspond to the stoichiometric condition, not ensuring to fulfill the increasing thickness condition. A thickness definition δ_r, comprising the reactive zone , i.e. the region where $\dot{\omega}_c \neq 0$, is thus proposed. This flame thickness is then computed on the preprocessing step for every individual one-dimensional flamelet, i.e. for each strain rate value, for both filtered and unfiltered conditions and the results are tabulated as $\delta_l^i = f(\widetilde{Z}, \widetilde{c})$.

3.2   Filtered flames characterization

The filtering of a flamelet has been expressed from the numerical point of view. This section depicts the effects of this operation on the flame structure. Two applications are addressed: the individual profile transformation and the computation of the model closure terms. The main features are described considering both an individual flamelet and the entire manifold. Subsequently special attention is paid on the species prediction and the flamelet displacement within the manifold.

3.2.1   General features

The FTACLES approach performs the filtering operation in the physical space, in opposition to a transformation in Z space as is the case for β-PDF integration. In the non-premixed case, this transforms the flame structure not only by modifying c as under premixed conditions, but also through Z filtering. In order to better understand the flamelet response to this numerical operation, Figure 3.1 presents the profile variation considering different filter sizes.

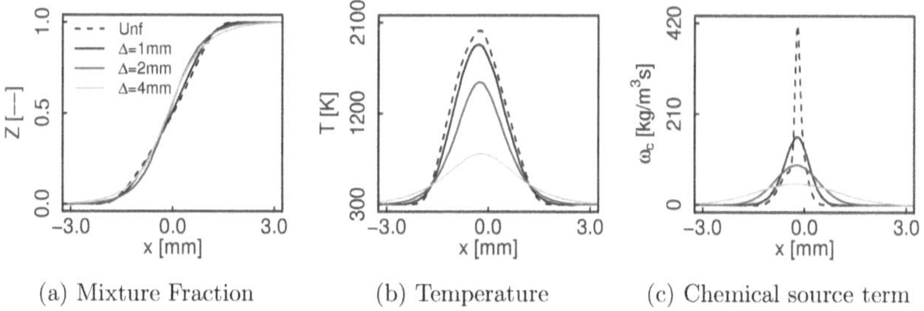

(a) Mixture Fraction   (b) Temperature   (c) Chemical source term

Figure 3.1: Profile transformation for a non-premixed CH_4-Air flamelet under three different filter sizes. Solid lines filtered profiles, dashed line the unfiltered reference.

One of the effects of the filtering process is to distribute the variable more smoothly along the domain, as in 3.1a, so that depending on the considered profile this might lead to the variable propagation at the extremes. The filtering as well decreases the peak values for non-monotonic functions, as observed in Figure 3.1b, where T_{max} is considerably diminished. The comparison of figures 3.1b and 3.1c highlights the profile thickness relevance. The filtering effect will therefore increase as the thickness diminishes.

Contrary to premixed flames, the flame thickness is not an intrinsic property of diffusion flames. This quantity will not only depend on the composition but will change as well either as a function of time, for an unsteady evolving diffusion flame, or as a function of the strain rate. Hence, depending on the specific aspect to be assessed flame thickness definitions can be proposed, being aware of the limited interpretation of this feature. Two definitions are introduced to quantify the flamelet width variation caused by the filtering process. The total thickness δ_Z considers the domain where $Z \in [0.01, 0.99]$, i.e. it covers both the reaction zone and the surrounding diffusion layers. The second one δ_{ω_c} comprises the region where $\omega_c > \omega_{c,Max} \cdot tol$, which can be interpreted as an estimation of the width of the reaction zone. Figure 3.2 depicts the filter effect employing both definitions, where a significant decrease in δ_i for increasing K can be appreciated along the manifold. For instance, the ratio $\delta_{K_{Min}}/\delta_{K_{Max}} \approx 76$ for δ_{ω_c} and 42 for δ_Z in the unfiltered condition, which explains the bigger filter effect with augmenting K, given a fixed Δ. Moreover, since by definition δ_{ω_c} is smaller than δ_Z, it will be more sensitive to the filtering.

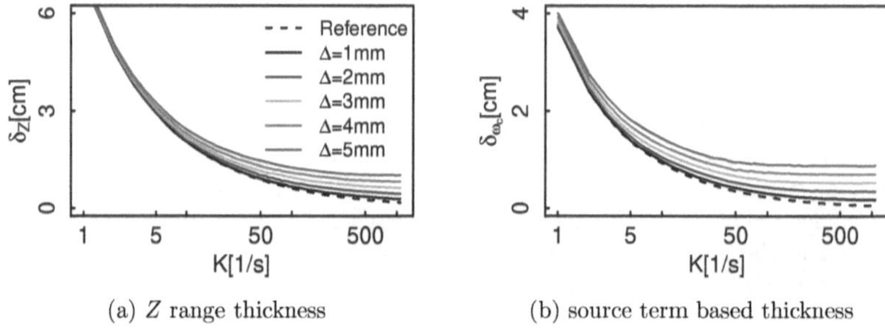

(a) Z range thickness   (b) source term based thickness

Figure 3.2: Flame thickness for a non-premixed CH_4-Air manifold.

It calls the attention that on both of the definitions there is a threshold K_{th}, after which the thickness converges for all the strain rates. To determine this threshold three flamelets have been taken, $K_1 = 1000s^{-1}$, $K_2 = 750s^{-1}$ and $K_3 = 500s^{-1}$, and their filtering evolution has been split as to better understand the transformation. Figure 3.3a presents a clear difference between the three unfiltered profiles. The three profiles continue to be distinguishable for $\Delta = 1$mm, where K_1 already undergoes a significant deformation, while K_3, due to its higher initial thickness appears to be less affected. As the filter size increases, the flamelet characteristic features vanish, it expands over the domain and its profile coincides with a lower strained flamelet. It is for this reason that the profiles are less distinguishable on Figure 3.3b, as K_1 and K_2 tend towards K_3. A further filter size increase augments the expansion, so that K_{th} decreases. The flat profile after $K = 500s^{-1}$ for $\Delta = 4$ and 5mm on Figure 3.2a indicates that the higher strained flamelets have all been expanded and thus became identical, as is the case on Figure 3.3c.

<hr>

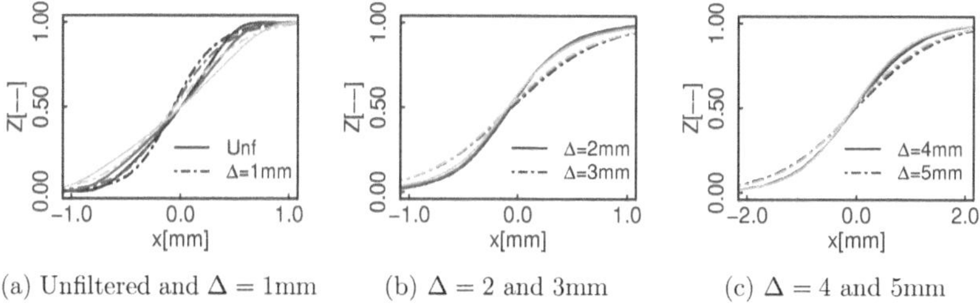

(a) Unfiltered and $\Delta = 1$mm   (b) $\Delta = 2$ and 3mm   (c) $\Delta = 4$ and 5mm

Figure 3.3: Mixture fraction profile transformation for three flamelets under different filter sizes. Black color $K_1 = 1000s^{-1}$, blue color $K_2 = 750s^{-1}$ and green color $K_3 = 500s^{-1}$.

Figure 3.2b showed that the above explained phenomenon holds as well for the second thickness definition, thus δ_{ω_c} converges for a given condition $K > K_{th}$. This does not necessarily imply that all the profiles coincide, but uniquely that they expand over the same length. Figure 3.4b compares each one of the filtered profiles against the unfiltered K_1. Even though K_1 unfiltered source term is approximately twice as big as the one corresponding to K_3, the difference significantly decreases for $\Delta = 1$mm. Figure 3.4c performs the comparison considering the filtered K_1 value at each Δ as the reference. The filter has a distinctive effect for each flamelet at $\Delta = 1$mm. From $\Delta = 2$mm the ratio becomes approximately constant and not equal to one. The results indicate that after a given filter size, in this case $\Delta = 2$mm, the three flamelets undergo a proportional transformation. The profiles expand over the same width, therefore δ_{ω_c} converges, though the individual peak values are not the same.

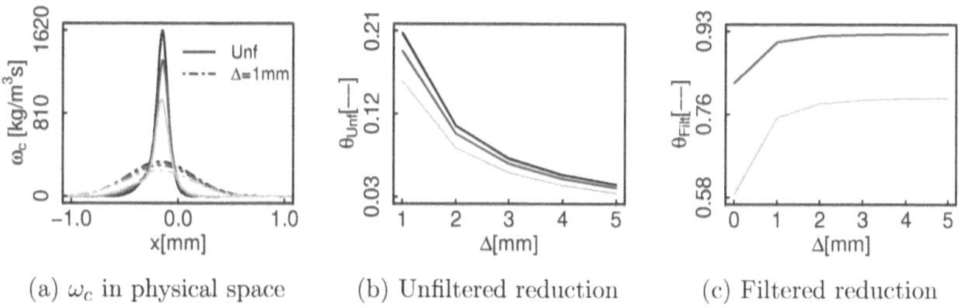

(a) ω_c in physical space   (b) Unfiltered reduction   (c) Filtered reduction

Figure 3.4: Progress variable chemical source term transformation for three flamelets under different filter sizes. Black color $K_1 = 1000s^{-1}$, blue color $K_2 = 750s^{-1}$ and green color $K_3 = 500s^{-1}$.

Expanding the analysis from three flamelets to the whole manifold, Figures 3.5a and 3.5b present the decrease in the maximum value for two non-monotonic variables, namely T and ω_c along a filtered manifold. Even though the $\omega_{c|K_{Max}}$ is around three orders of magnitude greater than $\omega_{c|K_{Min}}$ for the unfiltered flamelets, at $\Delta = 5mm$ the difference between both is not bigger than one order of magnitude. A substantial modification takes place at K_{Max} while at K_{Min} almost no change is observed. This reduction is accompanied by a propagation of the source term over

Chapter 3

the domain, as shown in figure 3.5c, which corresponds to the projection of δ_{ω_c} in Z space. This behavior increases with the filter size and K till a certain threshold, after which a non-zero source term is to be found all over the domain.

The great variation of the flame width and the consequent filtering effect along the manifold highlight the need of a three-dimensional space parametrization as $\varphi = f(Z, c, \Delta)$. In order to appraise the capability of the controlling variables to uniquely describe any point in space two tests are performed. The first one considers the evolution of different flamelets, i.e. varying K, given a fixed filter size. Figure 3.6a shows non-overlapping profiles, thus an adequate parametrization. The second scenario corresponds to fix K and evaluate the flamelet response for different Δ. Each filtered profile is clearly identifiable on Figure 3.6b, which demonstrates that the uniqueness condition is kept as well.

3.2.2   Correction terms computation

There are two applications of the filtering operation: the first one has already been addressed and corresponds to the transformations in the flame structure, whose result is a modified filtered manifold. The second application concerns the filtering of the parameterizing variables transport equation budget terms. The non-premixed FTACLES utilizes this approach to compute the unresolved terms on the LES context. The correction terms computation can be described as follows:

1. The flamelet is read and each one of budget terms of the transport equation, i.e. diffusive and convective fluxes, both in axial and radial directions, are computed.

2. The terms are filtered, i.e. they correspond to the resolved scenario.

3. The flamelet is filtered and the budget terms are computed. This corresponds to the approximation, or LES solution.

4. The difference between the resolved and the approximated terms is calculated and stored in the table as $f(Z, c, \Delta)$.

Figure 3.7a exemplifies the budget terms both for the resolved and the approximated solution, i.e. the outcome of steps 2 and 3. Figure 3.7b presents the actual tabulated data resulting from step 4, where the summation of the radial and axial convective correction terms gives the total convective correction $Conv_{tot}$.

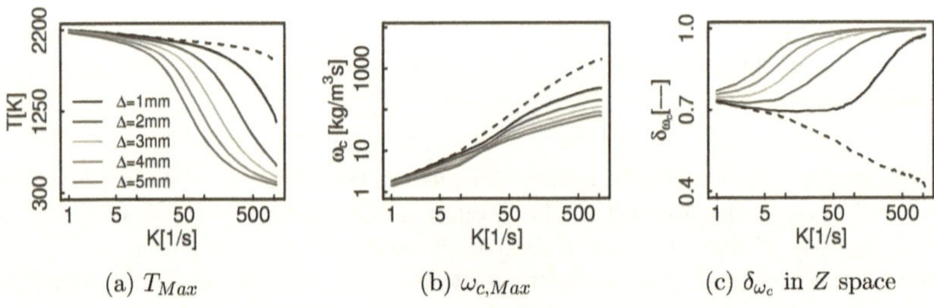

(a) T_{Max} (b) $\omega_{c,Max}$ (c) δ_{ω_c} in Z space

Figure 3.5: Maximum T and ω_c evolution along a filtered manifold. Solid lines filtered profiles, dashed line the unfiltered reference.

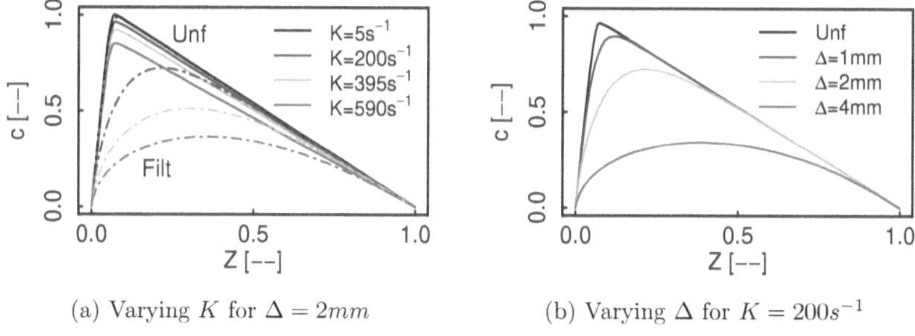

(a) Varying K for $\Delta = 2mm$ (b) Varying Δ for $K = 200s^{-1}$

Figure 3.6: Manifold parametrization uniqueness for (a) fixed Δ and (b) fixed K.

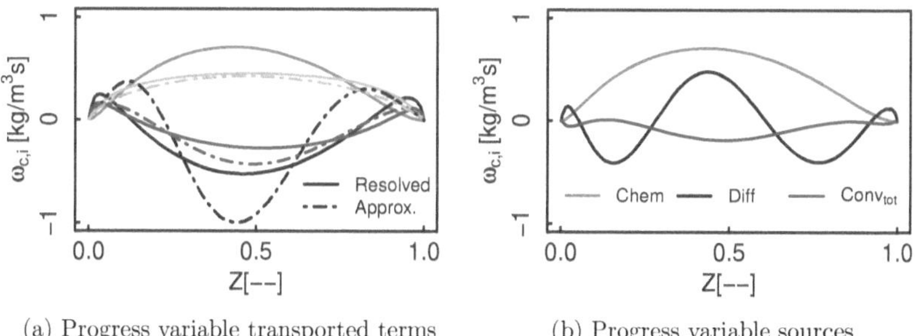

(a) Progress variable transported terms   (b) Progress variable sources

Figure 3.7: Progress variable transported and source terms in Z space, for a non-premixed $CH_4 - Air$ flamelet with $K = 200s^{-1}$ and $\Delta = 2mm$. Figure 3.7a red color chemical term, black color diffusive term, blue color axial convection, green color radial convection.

3.2.3 Retrieved variables prediction

One specific effect of the filtering operation is to alter the species prediction. The degree of modification for a given Δ depends on the considered species contour shape and on its thickness. For instance Figure 3.8 presents CO_2 transformations which preserve the original shape, in opposition to the response delivered by CO and OH that undergoes non negligible changes.

It is well known that when employing flamelet approaches the correctness in the species retrieval will be highly influenced by c definition. However the above shown effect exclusively depends on the filtering operation and the modification of $Z - Y_i$ relation. Therefore the deviations in CO profile become an intrinsic feature, which can not be altered by eventually incorporating this species into c. For other minor species, which tend to be localized in specific flame regions, i.e. OH, their behavior is altered as well, resulting in profiles which spread all over the space. Figure 3.9 shows the deviation with respect to the reference profile for each one of the three filters.

The ε_{CO_2} increases with the filter size and reaches its maximum value close to CO_2^{Max}. The profile extension towards $Z = 0$ causes CO reference and filtered

---

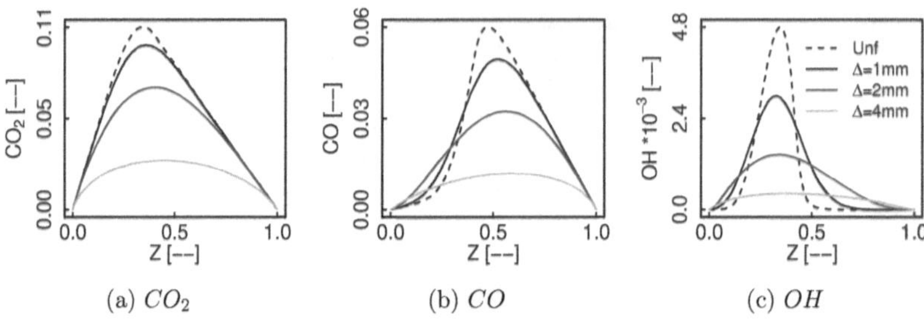

(a) CO_2 (b) CO (c) OH

Figure 3.8: Species profile transformation for a non-premixed CH_4-Air flamelet under three different filter sizes. Solid lines filtered profiles, dashed line the unfiltered reference.

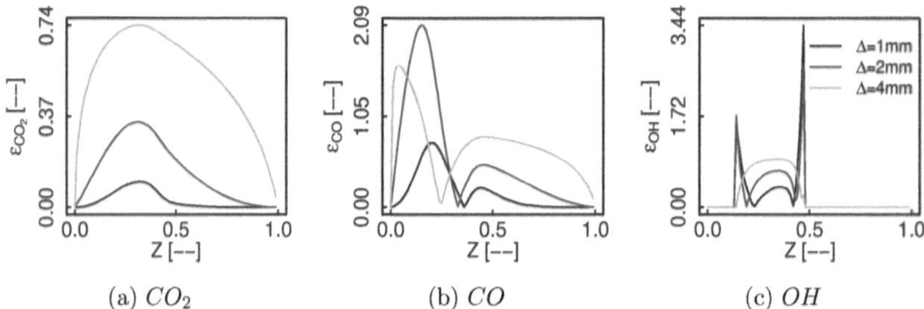

(a) CO_2 (b) CO (c) OH

Figure 3.9: Species profile deviation with respect to the unfiltered reference flamelet for a non-premixed CH_4-Air flamelet under three different filter sizes.

curves to intersect, so that ε_{CO} behavior as a function of Δ presents two peaks. The ε_{OH} has been truncated there where the reference reached zero, while for the rest of the domain it increases directly with Δ.

The individual species relevance on the filter effect can be further appraised in Figure 3.10 where they have been depicted as a function of $c = Y_{CO_2} + Y_{CO}$.

The filtering operation tends to homogenize the variable's behavior and vanishes individual traits so that all of them finally resemble. The lower row presents the normalization of each species based on the unfiltered profile and subsequently scaled considering the respective filtered c. The three observed species contours vary their amplitude, becoming significantly slender. CO and OH profiles lose their irregularity and tend to symmetrically evolve around the line defining its maximum. The decrease in the normalized peak values, i.e. the deviation from the identity, highlights the profile thickness filtering sensitivity. Thus while CO_2 depicts an almost linear correlation with c, CO shows a more pronounced decrease, and this effect is further amplified for OH. This is a crucial issue, as it implies that the retrieved variables mean value deviation as well as the fluctuations will vary for each observed species. Concretely, these results point out that for CO_2 the mean and RMS values prediction will be fully determined by c behavior, while for cases like CO and OH an additional discrepancy should be taken into consideration.

The analysis is valid as well for the thermophysical properties, whose thickness

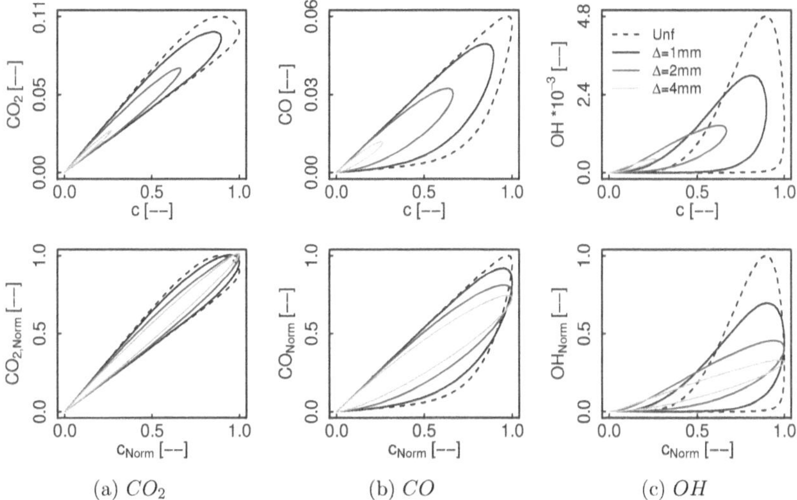

(a) CO_2 (b) CO (c) OH

Figure 3.10: Species profile transformation for a non-premixed CH_4-Air flamelet under three different filter sizes as function of the progress variable. Lower row corresponds to the normalization with $c_{i,max}$.

might not necessarily coincide with that one of c. Figure 3.11 shows an almost linear correlation between T and c, together with a profile narrowing in the normalized representation. The filter exerts a lower impact over the temperature profile than it does over c, therefore T_{Norm} values exceed unity, and this effect increases with Δ.

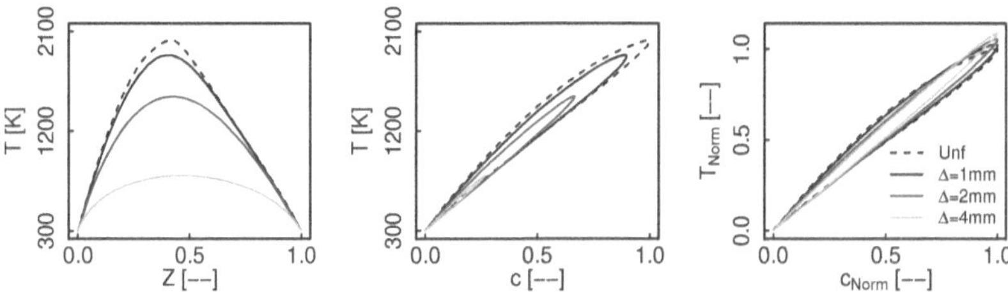

Figure 3.11: Temperature profile transformation for a non-premixed CH_4-Air flamelet under three different filter sizes as function of Z, c and c_{sc}.

The same filtering operation should be applied to all the retrieved values to guarantee that the progress variable appropriately describes the flame structure. For instance, the harmonizing evolution shown in Figure 3.12a opposes to figures 3.12b and 3.12c, where the use of distinct operators drastically distorts the flame structure. The results emphasize the lack of consistency when uniquely filtering the transported parameters and not the retrieved variables.

One of the main objectives of turbulent combustion models using a selected progress variable is the estimation of $\dot{\omega}_c$, the species and thermophysical properties

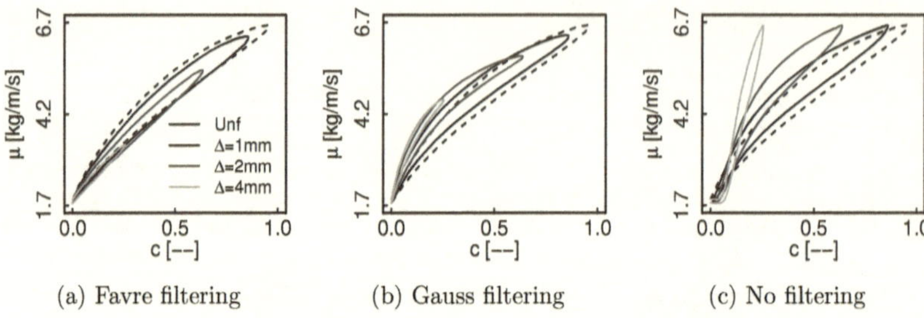

(a) Favre filtering   (b) Gauss filtering   (c) No filtering

Figure 3.12: Dynamic viscosity profile employing Favre filtering (a), Gauss filtering (b) and no filtering (c) as function of the Favre averaged progress variable.

prediction being conditioned to an adequate c computation. Figure 3.13a highlights ω_c unlikeness with respect to c. The unfiltered profile remains equal to zero for the lowest c range and it abruptly increases at the highest c values. The term steepness induces a particular response to the filtering operation which bears no resemblance neither with c nor with the rest of retrieved variables, as made evident by Figure 3.13b. The filtered profile undergoes a severe transformation at $\Delta = 1$mm, and for $\Delta = 2$mm it has been completely distributed along the domain so that the original features can be no longer recognizable. Figure 3.13c computes the filtered profiles deviation with respect to the unfiltered term. Close to the maximum value ε_{ω_c} increases with Δ, and it explodes at the extremes as the profile is extended.

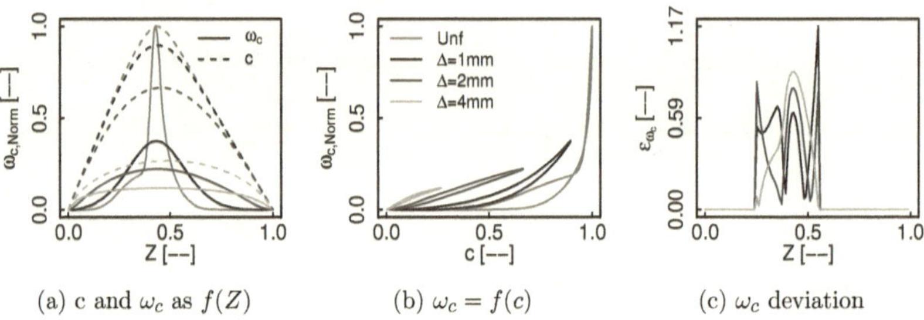

(a) c and ω_c as $f(Z)$ (b) $\omega_c = f(c)$ (c) ω_c deviation

Figure 3.13: Normalized ω_c profile in Z(a) and c(b) space, and the deviation with respect to the unfiltered flamelet(c) using three different filter sizes.

3.2.4   Flamelet displacement

The modification of the progress variable trajectories as function of the mixture fraction is a characteristic feature of non-premixed flamelet filtering. It causes a flamelet displacement as observed on Figure 3.6a. Aiming to further understand the mechanism as well as the implications of this phenomenon, a set of manifolds, one unfiltered, and four filtered with $\Delta = 1, 2, 3, 4$mm were created. The flamelet passing through an arbitrary point $P_0 = (Z_0, c_0)$ was then identified for each one of the manifolds. These flamelets trajectories were depicted for different retrieved variables, thus the predictions on the point P_0 were compared.

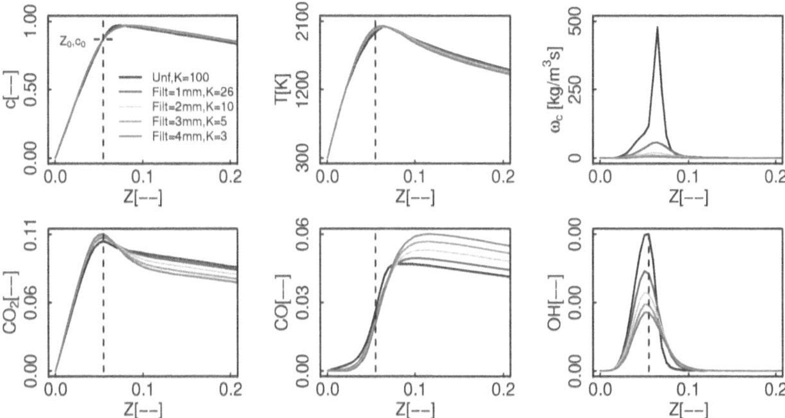

Figure 3.14: Filtered flamelet displacement and effect over the retrieved variables for different filter sizes.

The strain rate significantly decreases with Δ, as the faster flamelets collapse towards the bottom of the manifold. Despite the significant K variation, the constraint of intersecting P_0 defines considerably similar c profiles. This will in general hold for low unfiltered K values, and discrepancies will be observed for more strained unfiltered trajectories. Due to its strong linear correlation with c, the different T profiles tend to converge, thus the filtered flamelets describe a very similar condition to the unfiltered trajectory. A diametrically opposed behavior can be observed for ω_c, as the term almost vanishes from $\Delta = 2$mm. This responds mostly to the flamelet displacement and not to the trimming of the chemical source profile itself, as the shifted less strained flamelets possess significantly lower ω_c. Even though for this case $c = CO + CO_2$, not only the individual species trajectories differ, but even the prediction at P_0 is clearly distinguishable for each flamelet. For the case of OH the profile undergoes a substantial decrease, but contrary to ω_c, this is the result of both a high filtering sensitivity, due to its low thickness, and of the flamelet shift effect.

Figure 3.15 presents the strain rate contour over the unfiltered manifold as well as over the biggest filtered manifold, i.e. $\Delta = 4$mm. Even though exclusively steady flamelets have been employed, the higher K flamelets undergo an enormous deformation, so that they cover almost all the chemical space.

Figure 3.16 presents the temperature and chemical source term contours for the same two conditions. The coincidence in T profile already acknowledged in Figure 3.14 can be recognized here as well, as the predictions in the high c region of the manifold strongly resemble the unfiltered ones. The chemical source term on the contrary has no similarity at all with its unfiltered counterpart. The unfiltered peak values, located close to the stoichiometric condition, decrease by two orders of magnitude and expand over a wide region, being therefore deprived from their original significance.

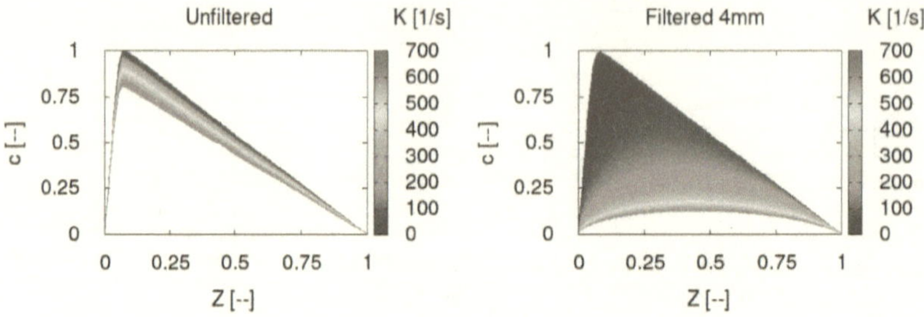

Figure 3.15: Mapping of K for an unfiltered and a filtered manifold with $\Delta = 4$mm.

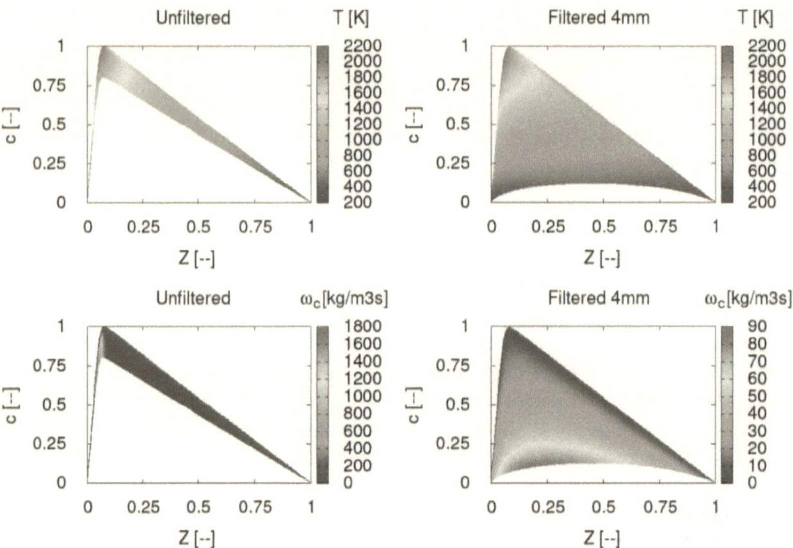

Figure 3.16: Mapping of T and ω_c for an unfiltered and a filtered manifold with $\Delta = 4$mm.

Figure 3.17 presents the decomposition of c source term into the filtered chemical part, the model correction and the resulting source term. The filtering operation moves the higher strained flamelets, i.e. those with higher chemical source term, to the bottom part of the manifold. Additionally the peak value is trimmed and the profiles extend over the domain. The correction term presents high positive values at a location coinciding with the most reactive region of an unfiltered manifold. This can be interpreted as a model compensation to the the filtering operation shifting the higher strained flamelets and consequently their chemical source terms, to the bottom of the manifold. It results that the total source term ω_{cTot} conserves the trend of the unfiltered chemical source term.

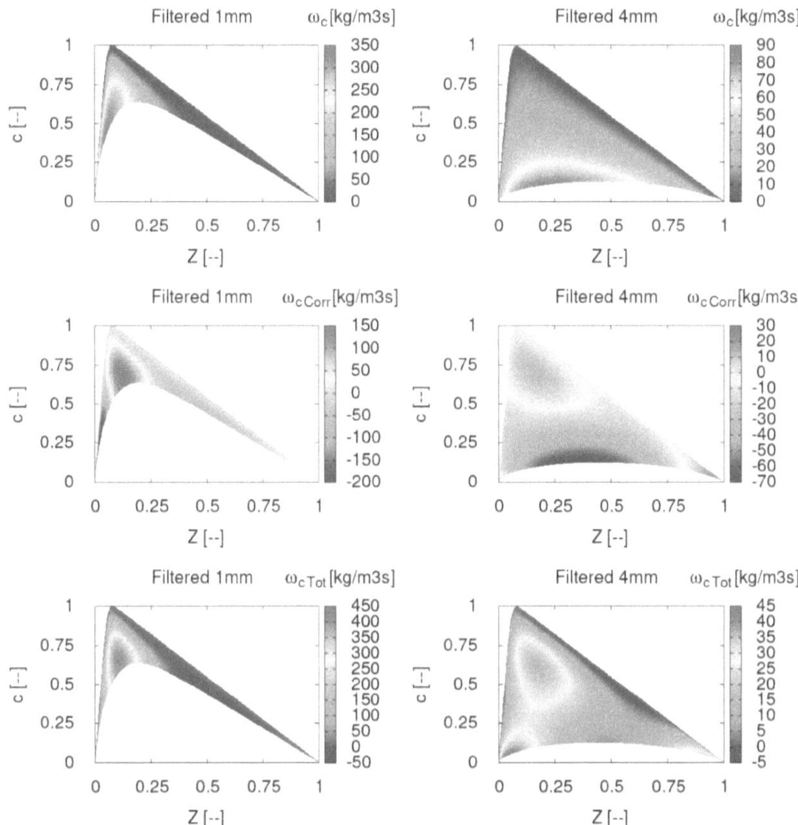

Figure 3.17: Non-premixed FTACLES progress variable source term decomposition: filtered chemical part (top), model correction (middle) and summation (bottom), for two different filter sizes.

3.2.5 Continuation methods

One particular case of the flamelet displacement is encountered when the filtered profiles fall in the region below the last steady ignited flamelet on the unfiltered manifold. The increase of a non-premixed flamelet strain rate decreases its residence time till a point where it is no longer enough for the reactions to be completed and the flame extinguishes. Consequently, a manifold employing steady counterflow diffusion flamelets covers the (Z, c) space from the equilibrium up to the extinction condition, and then it abruptly moves to the pure mixing line, thus leaving an empty region between these two conditions. In order to solve this situation unsteady flamelets can be used as proposed in [122]. An ignited flamelet is taken as initial solution, the strain rate is set to a value above extinction, and the flamelet's time evolution is tracked.

Since in this approach the additional dt term is not considered or tabulated, this works as a continuation for the properties from the quenching condition till pure mixing, analogous to an interpolation method. Indeed, this time-dependent condition prevents the application of the FTACLES formalism on the unsteady flamelets. The base assumption for applying the correction terms is to satisfy the transport equation, which is not possible if an existing time dependent term is neglected. As a result of that, the unsteady flamelets do undergo a filtering operation in order to preserve the uniqueness condition in the parametrization, but the additional model correction terms are neglected.

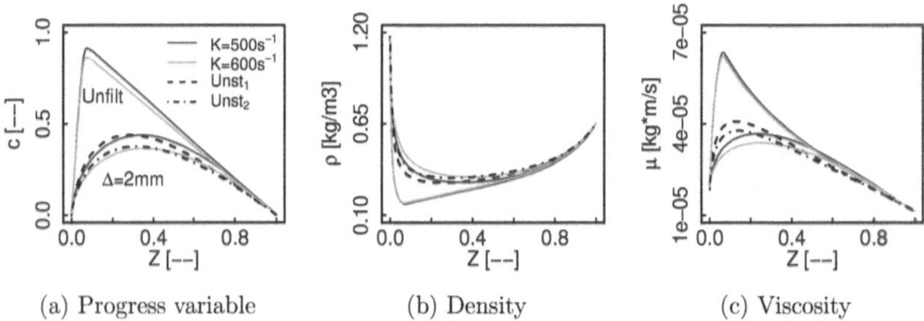

(a) Progress variable   (b) Density   (c) Viscosity

Figure 3.18: Progress variable, density and viscosity behavior for equivalent trajectory unsteady and filtered steady flamelets for $\Delta = 2$mm.

As already described in the previous section, the filtering operation modifies c trajectories, shifting them downwards in the manifold. It results that ignited flamelets start covering the space that under unfiltered conditions would correspond to unsteady flamelets. In order to further comprehend this transformation, two steady flamelets have been filtered and their profiles were compared against unsteady flamelets with similar c trajectories in Figure 3.18. The c profiles match quite well all along the mixture fraction space, and the same can be stated for the properties coupling the flame structure and the flow field, namely, density and viscosity, with the biggest differences observed for μ.

Even though the interchangeability of the observed curves suggests that these filtered flames might as well play a continuation role, intrinsic differences regarding other flamelet features make the problem much more challenging. For instance, the interpretation of the chemical source term is not straightforward. For unsteady flamelets the time varying contribution is not taken into account, which means that c transported equation on the CFD will not be able to reach any steady condition for these points. Highly strained flamelets on the contrary possess high valued but considerably thin source terms, which makes them considerably sensitive to the filtering operation.

Figure 3.19 presents the source terms for both the unsteady as well as the filtered and unfiltered steady flamelets. There is a significant trimming on the peak value for the steady flamelets, leading to a reduction of factor 20, after which the profiles extend with a nearly constant value all over the domain. The unsteady flamelets source term do not share the order of magnitude of the unfiltered steady, constraining the use of logarithmic scale in order to appreciate them. Though the filtered steady flamelets present a difference with respect to the unsteady flamelets, the impact is rather small as both of the values are significantly low. These results

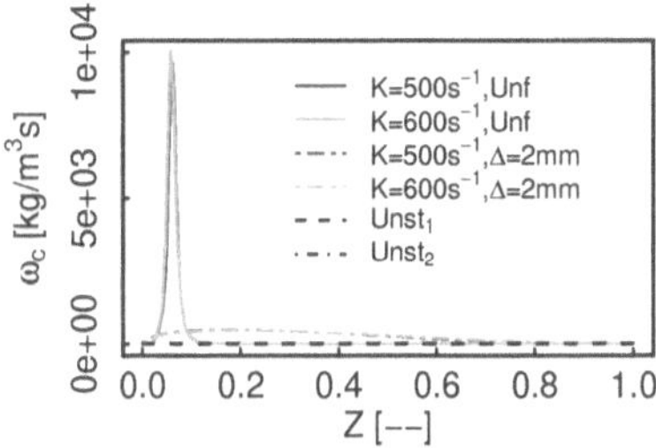
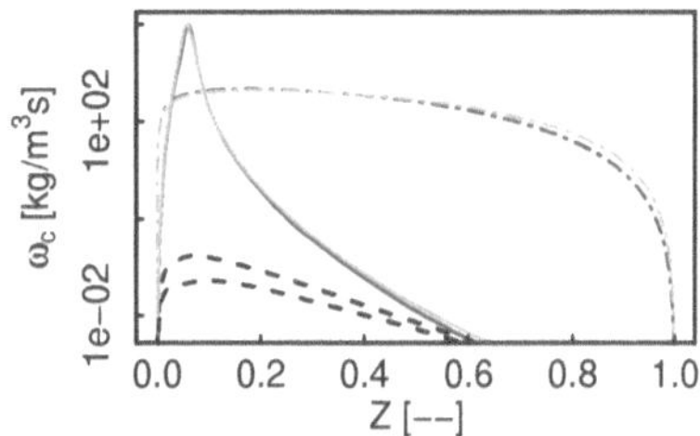

Figure 3.19: Chemical source term behavior for equivalent trajectory unsteady and filtered steady flamelets for $\Delta = 2$mm, plotted in normal and logarithmic scale.

are particularly encouraging as they indicate the physical coherency of the filtering operation. The filtered high strained flamelets and the unfiltered unsteady flamelets perform the same function, they serve for continuation purposes towards the pure mixing condition.

3.3 Premixed vs non-premixed FTACLES

Despite employing the same filtering principle, two substantial differences are worthy to be mentioned between the premixed and non-premixed FTACLES: the first one obeys topological reasons, while the second one is of theoretical nature. The filtering operation not only extends and smoothens a variable profile, but for non-monotonic functions it additionally decreases the peak value depending on the gradients and the filter size. Due to their different internal structure, many variables which might present a monotonic behavior in freely propagating premixed flames, e.g. c, CO_2 and T, will not on the counterflow diffusion counterpart. This severely affects the filtered profiles and might therefore as well impact the model results.

Concerning the theoretical foundation, an essential requirement during the premixed thickening operation, whether by means of a factor or by direct filtering, is the conservation of the laminar flame speed s_l^0. This condition cannot be applied to a diffusion flame since by definition the flame front does not propagate. It follows that while a diffusion flame front filtering operation can be clearly formulated from the numerical point of view, the conservation of a given property characterizing the flame is considerably more difficult to demonstrate. The goal of this section is to further comprehend the filtering operation effect in both regimes. In order to do this first the main features of premixed FTACLES are presented, and subsequently the output of both manifolds are compared. The section concludes summarizing the main distinctive features between the premixed and non-premixed FTACLES approaches and their implications.

3.3.1 Filter effect in premixed

In analogy to the comparisons performed on Figures 3.2 and 3.5 this section starts by describing the behavior of the characteristic length and velocity scales along a premixed manifold. Subsequently, as already done for the non-premixed regime, the filtering effects are appraised in two different ways: first fixing the filter size and applying the operation over different flamelets, then, fixing the flamelet and changing the filter sizes.

Similar to the ATF method, the FTACLES model changes the flame structure by increasing the flame thickness and decreasing the chemical source term. Figure 3.20 shows the laminar flame thickness, maximum chemical source term and laminar flame speed for a CH_4-air manifold, with lean to rich varying compositions, subject to different filter sizes. The unfiltered flame thickness profile reaches its minimum value close to the stoichiometric condition, i.e. $\theta = 1$, and increases towards the sides, while both $\omega_{c,Max}$ and s_l^0 present the inverse behavior. The flame thickness determines the flamelet sensitivity to the filtering operation, thus the biggest transformations take place close to stoichiometry. Figure 3.20b depicts $\omega_{c,Max}$ variation. For $\theta = 1$ the decrease is approximately of one order of magnitude for $\Delta = 1mm$ and of around two orders of magnitude for $\Delta = 5mm$, while the filter effect significantly decreases when moving towards the very lean and very rich regions.

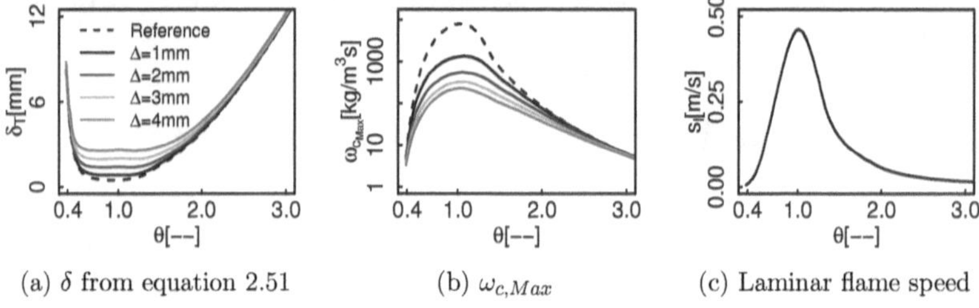

(a) δ from equation 2.51 (b) $\omega_{c,Max}$ (c) Laminar flame speed

Figure 3.20: Premixed FTACLES effect on the flame thickness, chemical source term and laminar flame speed for a CH_4-air manifold.

Three flamelets with compositions varying from almost stoichiometric towards a rich condition and filtered with $\Delta = 2mm$ are presented in figure 3.21. For better visualization the profiles have been shifted from their original position as not to superimpose. The variation in the flame thickness as a function of the composition can be appreciated through the decrease in c slope as the equivalence ratio increases. This explains the fact that the filter exerts very little effect for $\theta = 1.61$, while a considerable change is seen instead for $\theta = 0.98$.

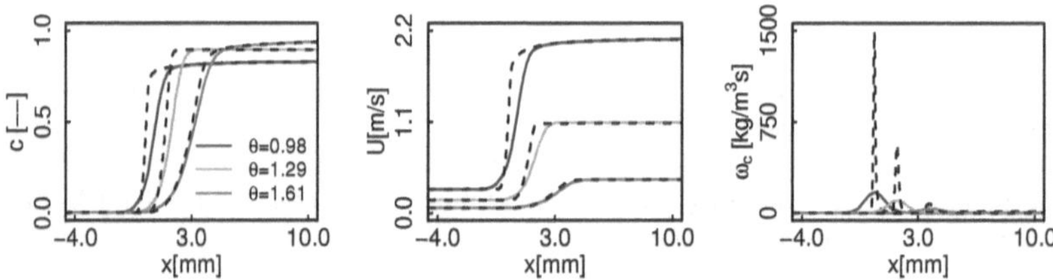

Figure 3.21: Premixed FTACLES profile transformation for different equivalence ratios and $\Delta = 2mm$.

The filtering does not only decrease ω_c peak value, but it also extends the profile over the domain. Figure 3.22 depicts a coordinate transformation into c space employing $\Delta = 1mm$, in order to better appreciate the profiles due to the source term higher filter sensitivity with respect to c or u. The source term is no longer confined to a specific region, and will indeed extend all over the domain after a threshold is surpassed. This is a substantial difference between the FTACLES and the ATF approaches, namely, in the latter a coefficient is applied to both the diffusivity and the source term, but the profile in c space is not altered. From figure 3.22a a significant extension can be seen for $\theta = 1.29$, while for $\theta = 0.98$ almost the entire domain is covered. The filtering effect as function of the composition can as well be appreciated in the magnitude of ω_{conv} as shown in figure 3.22b. For instance at $\theta = 0.98$ both terms are in the same order of magnitude as can be seen in 3.22c.

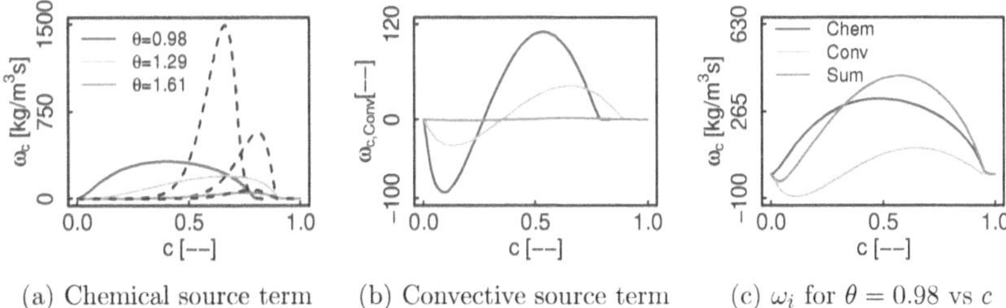

(a) Chemical source term   (b) Convective source term   (c) ω_i for $\theta = 0.98$ vs c

Figure 3.22: Premixed FTACLES source term behavior in c space for different equivalence ratios and $\Delta = 1mm$.

A second scenario consisting on freezing the composition and varying the filter size is shown in Figure 3.23 for $\theta = 0.98$. It results important to recall the complete change in the shape of ω_c. As the filter size increases the variable is distributed all over the domain and the profile becomes quite homogeneous.

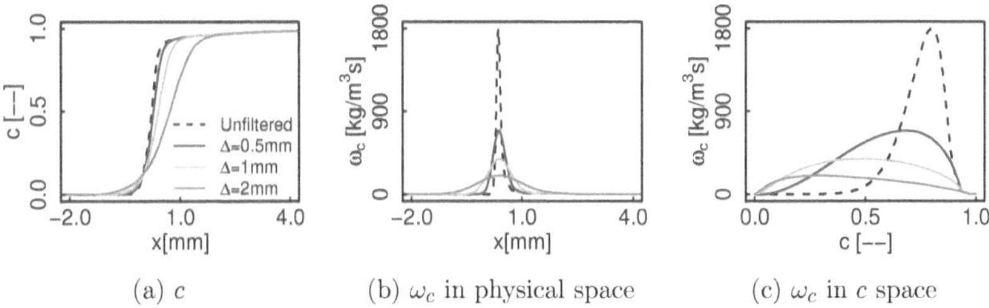

(a) c (b) ω_c in physical space   (c) ω_c in c space

Figure 3.23: Premixed FTACLES profile transformation with $\theta = 0.98$ and different filter sizes.

3.3.2 Manifold transformation

Once having described the main features on premixed flamelet filtering, this section assesses the filtered outcome for both premixed and non-premixed regimes. Since premixed and non-premixed flamelets evolve in different directions in chemical space, first the reference flamelet should be chosen. One option is to select a non-premixed flamelet and then extract the corresponding (Z, c) trajectory on the premixed manifold. Another strategy was used here, namely, to fix the premixed flamelet and compare it against the corresponding composition iso-Z contours extracted from a non premixed manifold. Three variables are analyzed, namely the viscosity, the chemical source term and CO.

Figure 3.24 presents the results for five different compositions. Only steady state diffusion flamelets have been considered, which explains the truncation of the non-premixed profiles. The first row presents the unfiltered manifolds, while the second and third depict the resulting filtered profiles with $\Delta = 2, 4$mm, i.e. the effect of the FTACLES approach on each combustion regime. The three chosen variables represent different situations: μ has a strong linear correlation with c, while ω_c and CO illustrate the relevance of the profile thickness for not linearly correlated variables.

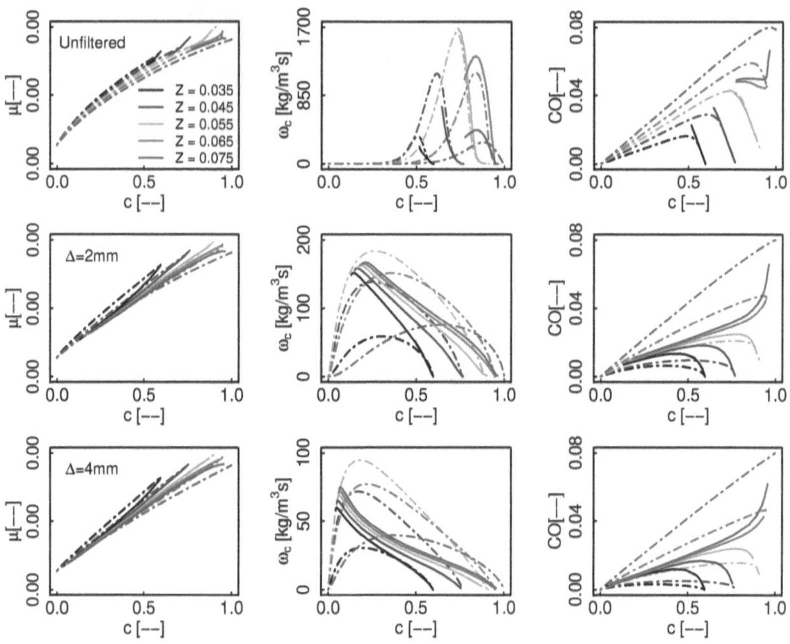

Figure 3.24: Comparison between premixed and non-premixed FTACLES profile transformation at different Z-isolines and filter sizes. Solid lines non-premixed, dashed lines premixed flamelets.

For the non-premixed case, ω_c uniformly spreads throughout the domain with increasing Δ and K, as previously highlighted in Figure 3.1c. Hence the non-premixed FTACLES iso-Z curves tend to converge as c decreases, despite the clearly distinguishable unfiltered profiles. This is coherent with the qualitative results of Figure 3.16. For the premixed case on the contrary, though ω_c peak decreases, and the profile extends all over c domain, each one of the iso-Z contours continue to be clearly identifiable. For the analyzed case the non-premixed model predicts a lower value for Z_{st}, and might over estimate as the Z distance from this point increases. Due to the higher reduction at stoichiometry, the non-premixed approach globally delivers smaller values than the premixed counterpart, thus appearing to be more sensitive to the filtering operation.

The unfiltered non-premixed straight CO profiles evolve into curves that resemble the premixed case, while a rather linear profile extension takes place for μ. This agrees with the statements made on Section 3.2.5, namely, the filtering operation works as a continuation method over the unsteady region in the manifold. However the farther the flamelet is driven from its original trajectory, the more uniform the retrieved profile becomes, e.g. the different profiles collapse. The comparison against premixed profiles indicates that special attention should be paid in the lower region of the manifold, i.e. small c values, as big errors might be introduced.

The flamelet displacement explained in section 3.2.4 can be appreciated as the lower c on the non-premixed trajectories, i.e. c_{min}, decreases with Δ. Figure 3.25 offers a clearer vision of the undergoing process, as it evidences the varying flamelet response along the manifold. The constant dK/dc in 3.25a shows that K variation between any arbitrary pair of flamelets originally produces an equivalent c decrease. This property does not hold anymore for the filtered manifold, as the profile distortion varies with the thickness, i.e. with K. Therefore a non-uniform evolution along c is obtained, whose trend holds for all the extracted iso-Z contours. Figure 3.25c presents a huge flamelet agglomeration on the low c region. The strain rate gradient starts decreasing around $c = 0.2$, which for instance corresponds to $K = 160s^{-1}$ for the trajectory at $Z = 0.055$ denoted with the point P_{55}. This has a direct practical consequence for the manifold construction, namely that different flamelet spacing is necessary in order to guarantee the proper resolution for varying filter sizes, and respect to the unfiltered manifold.

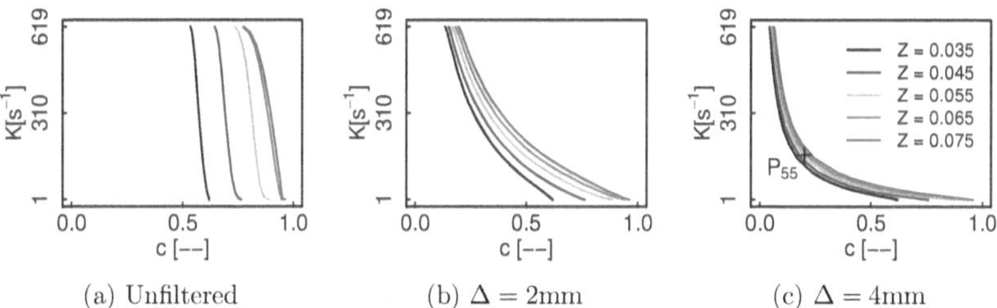

(a) Unfiltered   (b) $\Delta = 2$mm   (c) $\Delta = 4$mm

Figure 3.25: Flamelet redistribution along the manifold at different Z-isolines and filter sizes.

Figure 3.26 shows the temperature contours for both the premixed and non-premixed manifolds, unfiltered and after applying the filtering operation with $\Delta = 4$mm. The differences on the premixed temperature field appear to be very subtle. On the premixed regime, the mixture fraction does not change within a flamelet and the only controlling parameter is the progress variable. Due to c monotonic evolution, the filtering operation extends the profile physically, while it remains unmodified in chemical space. This is a fundamental issue, then it indicates that a given filtered species transformation in chemical space will be determined by its degree of correlation to c. For instance, c and T are highly correlated, therefore the temperature distribution stays unaltered in the filtered manifold.

For the non-premixed case, on the contrary, a significant effect can be appreciated all over the manifold. The flame structure on a counterflow diffusion flamelet is described in terms of the mixture fraction. It follows that the filtering effect in chemical space will depend on the variable's correlation to Z. However, in order to take into account the non-equilibrium effects, a second parameterizing variable, e.g. c, becomes necessary. The progress variable evolves non-monotonically along Z, therefore its filtered profile is not only extended in physical space but also its maximum value decreases. The filtering operation changes its relation to Z, hence c representation in chemical space. This elucidates the considerable T transformation, characterized by a lower maximum value distributed over a wider region.

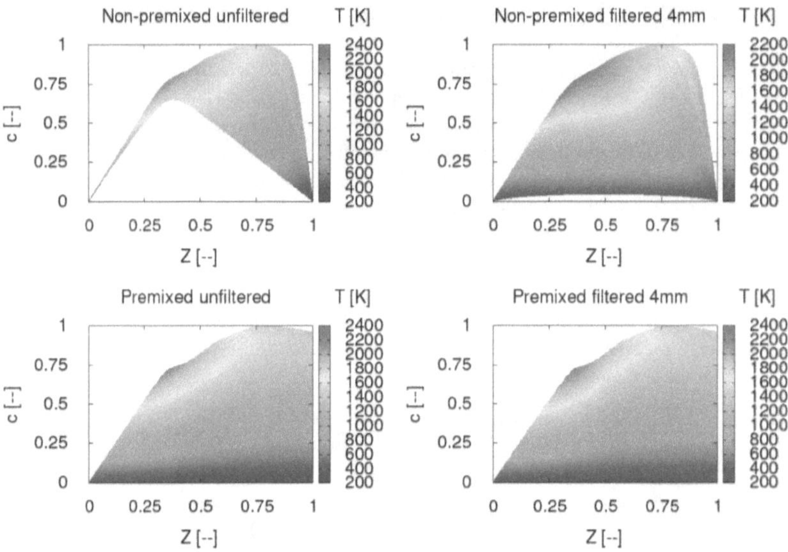

Figure 3.26: Mapping of the temperature for non-premixed and premixed manifolds, unfiltered and filtered with $\Delta = 4$mm.

Figure 3.27 shows the transformation for ω_c. The directionality of the filtering operation, moving along the Z direction for the non-premixed and at iso-Z curves for the premixed, can be clearly recognized. Thus, while the maximum value remains in the same order of magnitude, the distribution is completely dissimilar. The entirely distinct response of the chemical source term with respect to that observed on the temperature obeys to each species profile thickness, as well as to their degree of correlation with c.

After introducing the main features of the filtering operation over premixed flamelets, this section showed clearly distinguishable manifold transformation features on the two regimes. The relation between $Z - c$ is altered on non-premixed flames, thus the retrieved variables behavior in chemical space is inevitably modified with increasing filter size. For the premixed condition on the contrary, c transformation impacts exclusively the physical space. Therefore, the filtered species deviation

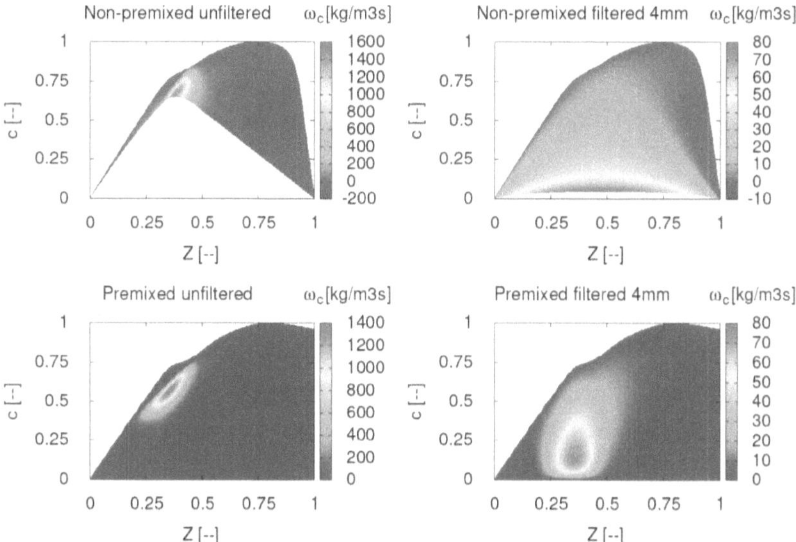

Figure 3.27: Mapping of ω_c for non-premixed and premixed manifolds, unfiltered and filtered with $\Delta = 4$mm.

will depend on their correlation to the progress variable.

3.3.3 Distinctive features

Section 3.3.2 clearly indicates that though numerically identical, the filtering operation exerts a completely different effect on premixed and non-premixed flamelets. Transformations on the retrieved variables were appreciated by considering $\varphi = f(Z, c)$ behavior either along iso-Z surfaces or throughout an entire manifold. This section summarizes the comparison between the premixed and non-premixed FTACLES approaches addressing two questions: what are the differences concerning the model assumptions? and what are the implications for the formalism?

Table 3.2: Distinctive features bewteen the premixed and non-premixed FTACLES.

	Assumption	Premixed	Non-premixed	Effect
Charact. velocity	$s_\Delta = s_l$ no SGS wrinkling	s_l	None	Lack of conserved property
Charact. thickness	$\delta < \Delta$	Flame property	$f(t)$ or $f(K)$ $\delta_d \approx \left(\frac{D}{K}\right)^{0.5}$	Uneven Δ sensitivity
c evolution		Monotonic	Non-monotonic	Flamelet shift

Concerning the theoretical foundation, an essential requirement during the premixed thickening operation, whether by means of a factor or by direct filtering, is the conservation of the laminar flame speed s_l^0. This condition cannot be applied to a diffusion flame since by definition the flame front does not propagate. Section 3.2.4 analyzed the flamelet displacement by fixing an arbitrary point $P_0 = (Z_0, c_0)$ and retrieving the values on different filtered manifolds. The methodology can be inverted to verify the correspondence between a filtered and an unfiltered trajectory within the manifold. In this case a flamelet label K_{Label}, defined as the strain rate at the oxidizer side of the unfiltered solution, can be used to identify the flamelet from which each filtered profile has been obtained. In a diffusive flame chemical reactions take place in a thin region around Z_{st}, beyond which fuel and oxidizer streams mix. Acknowledging that the total flame extent $\Delta Z = [0, 1]$ in physical space will largely vary as a function of the strain rate, the flamelet identification is done considering $K_{Label}|Z = Z_{st}$. Hence the flamelet label will be employed to assess the capability of the non-premixed FTACLES formalism to conserve the flame structure on Chapter 4, and so substantiate the formalism consistency.

Premixed flames typical flame thickness varies in the range $[0.1 - 1]$mm [29] [147], thus it cannot be resolved on typical LES numerical grids. The premixed FTACLES exploits this flame property, and emphasizes the advantages of the approach to retrieve the correct laminar flame speed when $\delta_l^0 < \Delta$. For the case of non-premixed flames the flame thickness is no longer a flame property but it varies either with time or with K for strained flamelets. The mixing layer thickness scales with $K^{-0.5}$, therefore covering a wide range of values as shown in Figure 3.2, which explains the non-premixed flamelet manifold highly uneven filter sensitivity shown in Figure 3.5. This might dramatically change the model response along the domain, e.g. the flame structure undergoes a significant transformation on the upstream high-strained region whilst the degree of modification is rather negligible downstream.

The final distinctive feature, though not stated as a requirement for the premixed FTACLES model, substantially modifies the interpretation of the non-premixed counterpart. The filtering operation not only extends and smoothens a variable profile, but for non-monotonic functions it additionally decreases the peak value depending on the gradients and the filter size. Due to their different internal structure, many variables which might present a monotonic behavior in freely propagating premixed flames, e.g. c, CO_2 and T, will not on the counterflow diffusion counterpart. This is particularly relevant for the progress variable, for its trimming leads to the flamelet displacement explained in Section 3.2.4 and then compared with the premixed behavior in Section 3.3.2. Practically, this transformation leads to modified retrieved variables whose behavior is fully consistent with the filtered flame structure, but whose comparison for instance with experimental data might be not a straightforward task.

3.4 Non-premixed FTACLES vrs β-PDF integration

The FTACLES approach deals with the unresolved flame front by directly filtering the one-dimensional flamelets. Another alternative to address the problem is following an statistical approach, e.g. employing either presumed or transported PDF's. Despite the fundamental phenomenological differences between the filtering and the PDF integration, both operations have in common that they substantially modify the flamelet profile from the pre-processing stage. This section considers the case where the presumed joint PDF consists of a β-PDF for the mixture fraction and a δ function for the progress variable. The manifold transformation is compared and an interpretation of the non-premixed FTACLES formalism is given.

Aiming to further comprehend the similarities as well as diverging features, three flamelets $K = 5s^{-1}$, $K = 200s^{-1}$ and $K = 1082s^{-1}$ with $\delta_5 = 13$mm, $\delta_{200} = 1.5$mm and $\delta_{1082} = 0.5$mm have been analyzed in order to assess the influence of the flame thickness. Figure 3.28 shows c profiles for the normalized $Z_v^2 \in [0, 1]$ in 0.1 intervals altogether with filtered flamelet profiles employing three different Δ. The integration effect is fundamentally the same for the three flamelets. Hence, for a given retrieved variable, e.g. c, the response is independent of K, which radically differs from the filtering operation. Thus while for $K = 5s^{-1}$ no modification is exerted by any of the considered filters, three clearly distinguishable trajectories are depicted for $K = 200s^{-1}$ and finally for $K = 1082s^{-1}$ the profile is strongly distorted and all the filtered trajectories converge towards a quite high Z_v^2.

While the PDF distribution unequivocally depends on the Z, Z_v^2 combination, the filtering operation might significantly vary as a result of the spatial variable distribution as already pointed out in Section 3.2.3. Aiming to further assess this phenomenon, Figure 3.29 depicts the effect of both operations over CO_2, CO and OH. Due to the lack of filtering sensitivity observed in 3.28a, $K = 5s^{-1}$ has been omitted. For the two remaining strain rates exclusively the Z_v^2 trajectories which more closely approach to the filtered counterparts are displayed.

On a PDF approach the flamelet mixture fraction is not modified during the integration and therefore the detailed chemistry parameterizing variable is the same as the transported variable, i.e. $Z_{fl} = \tilde{Z}$. This is not the case for the filtered approach as the variable represented on the abscissa axis after filtering corresponds to $\tilde{Z} \neq Z_{fl}$. Moreover the filtering operation is performed in physical space while the integration is done on the mixture fraction counterpart. The awareness of the above mentioned features gives a particular relevance to the presence of corresponding Z_v^2-Δ pairs where the flamelet transformations appear to be considerably similar for the three appraised species.

The accordance between these results suggests the possibility of interpreting the filtered profiles in an analogous relation with the PDF integration, where the

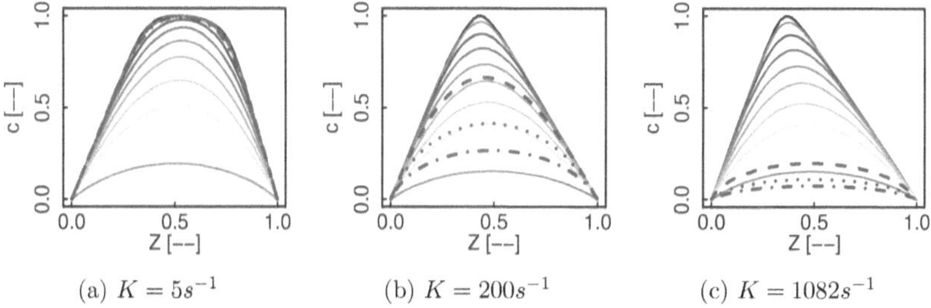

(a) $K = 5s^{-1}$ (b) $K = 200s^{-1}$ (c) $K = 1082s^{-1}$

Figure 3.28: Progress variable profile transformation due to β-PDF integration and direct filtering. Solid lines integrated profiles in $Z_v^2 = 0.1$ increase, blue dashed line $\Delta = 2$mm, blue dotted line $\Delta = 3$mm and blue dashed-dotted line $\Delta = 4$mm.

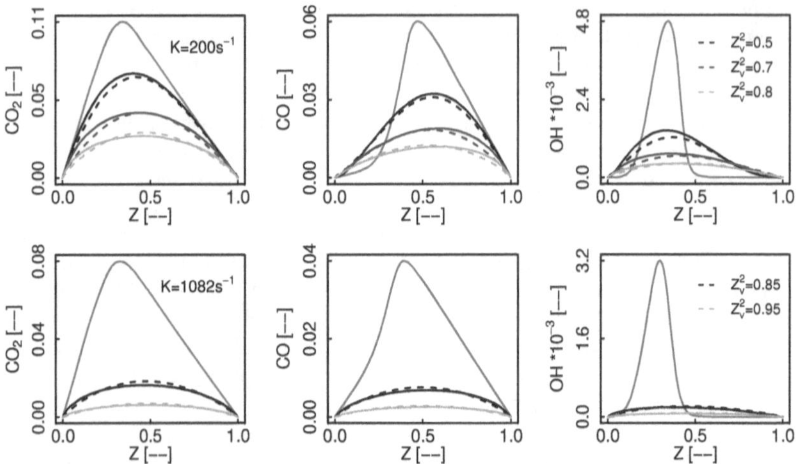

Figure 3.29: Species profile transformation due to β-PDF integration and direct filtering for two different flamelets. Red solid line original flamelet, black solid line $\Delta = 2$mm, blue solid line $\Delta = 3$mm, green solid line $\Delta = 4$mm, dashed lines β-PDF integration.

equivalent Z_v^2 directly increases with K for any given Δ. Thus from Figure 3.29, $\Delta = 2$mm is equivalent to integrate $K = 200s^{-1}$ profile considering $Z_v^2 = 0.5$, whilst for $K = 1082s^{-1}$ the same filter size corresponds to $Z_v^2 = 0.85$. In other words, the filtering approach induces huge reductions as K increases, which on the integrated approach is only obtained under considerably high fluctuation level, i.e. a severely distorted distribution. Nonetheless the underlying reason for this modification is absolutely different, as in the integrated approach it depends on the level of turbulence, while in the filtering approach it is uniquely mesh dependent.

3.5   CFD implementation

Two test cases were designed to verify the code implementation. They represent a filtered steady counterflow flamelet, and distinguish from each other by the numerical grid and table dimensionality. Table 3.3 summarizes the main characteristics.

The case set up can be described as follows:

1. Define K and Δ.

2. Compute either a one or two-dimensional filtered manifold.

3. Define two of the following parameters and compute the third one: Z_{in}, Z_{out}, or the domain length L.

4. Compute the boundary conditions as a function of Z or x.

5. Define Δ/h ratio and generate the numerical grid.

Table 3.3: Verification cases.

Case	Test 1	Test 2
Target feature	solver-library communication and table access	look-up algorithm
Grid dimensionality	1-D	2-D
Table dimensionality	1-D	2-D
Table parameters	Z	Z,c
Transported variables	U,p,Z	U,p,Z,c
Source terms	Z,p	Z,c

The results are compared against the direct filtering of a counterflow flamelet. Each one of the cases permits the assessment of specific issues, therefore a more detailed description is presented subsequently.

3.5.1 One-Dimensional Case

The goal of this test is to verify the table access and the solver-library communication. The correction terms computation has already been corroborated during the manifold construction considering the approximated budget terms. Therefore the model implementation is appraised comparing the mixture fraction and velocity profiles against a directly filtered one-dimensional flamelet. The correct spatial evolution of the transported variables portraits a successful filtering operation transformation.

In view of its dimensionality, the case requires just a few seconds to be resolved. Thus, it permits a quick debugging of the solver and the combustion library, i.e. the table reader and the combustion model. Due to the intrinsic two dimensional nature of diffusion flames, the radial convective term is added as a source term to satisfy continuity. Recalling that a pressure-based solver is used, this correction term is included in the pressure equation as pointed out in Table 3.3. A one dimensional table as a function of Z is employed, i.e. only one flamelet is used for the table generation and the closure terms are computed for a specific filter size.

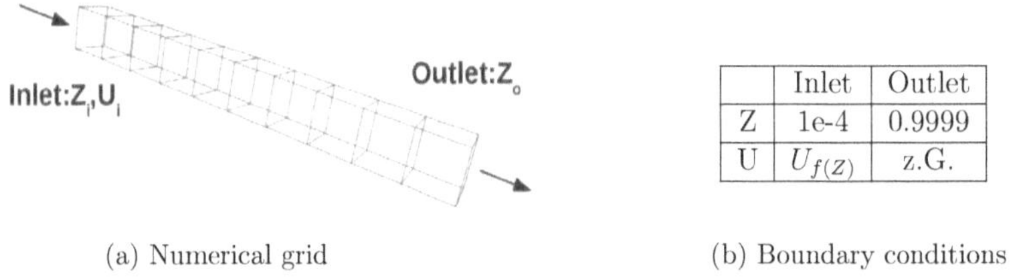

	Inlet	Outlet
Z	1e-4	0.9999
U	$U_{f(Z)}$	z.G.

(a) Numerical grid   (b) Boundary conditions

Figure 3.30: One dimensional case. z.G. stands for zero-gradient.

The numerical exercise was performed varying the strain rate, the filter size and the mesh resolution Δ/h ratio. Figure 3.31 presents the results for two K values and a filter size of 0.5mm. The model results coincide with the directly filtered detailed chemistry flamelet. The mixture fraction profile expands along the physical domain, the mass conservation is fulfilled, and the velocity field smoothens and extends. The first two variables are explicitly altered by the model correction terms, while the latter is implicitly transformed through the thermophysical properties.

3.5.2   Two-Dimensional Case

The second test assesses the adequate performance of the table look-up algorithm. A two dimensional table $\varphi = f(Z, c)$ with a fixed filter size is employed. The strain rate is first defined and subsequently the domain length and boundary conditions are set to match this specific flamelet. Figure 3.32 illustrates the goal of the test, namely, to identify and retrieve the correct trajectory from the manifold.

The two dimensional test case consists of a counterflow flame solved on an axisymmetric domain. It exploits the fact that a counterflow flamelet can be considered as a two dimensional problem if the radial term emerging from K is independently solved.

Figure 3.34 presents the progress variable profiles both in physical and in mixture fraction space. Two aspects can be observed, on one side the correct spatial evolution, i.e. the right profile displacement and peak trimming, and on the other the adequate Z, c coupling. The simulation results lie over the reference curves, and the numerical grid resolution is able to handle the scalar gradients.

The left column of Figure 3.35 presents the retrieved K profiles. Since the goal of the exercise is to locate a specific flamelet within the manifold, this variable indicates the precision with which the model approaches the reference. The results are under predicted on the oxidizer side, they increase towards the center of the domain and decrease once again towards the fuel side. At the manifold extremes all the flamelets converge, therefore a deviation on K profile has a rather small repercussion on the scalar prediction.

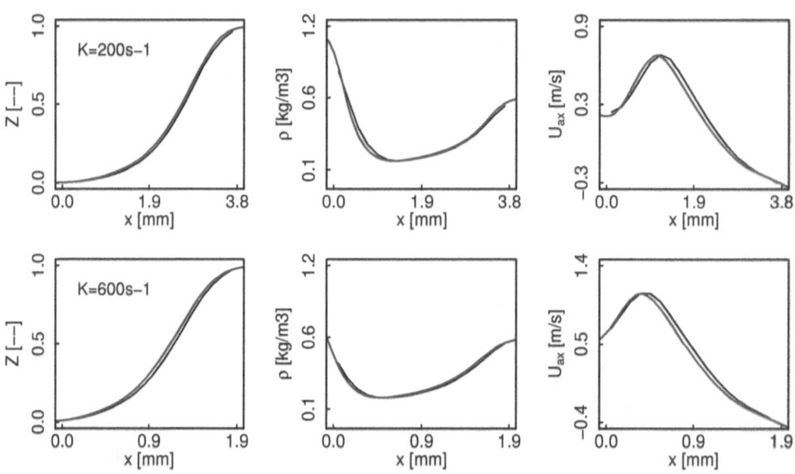

Figure 3.31: Filtered mixture fraction, density and velocity profiles with $\Delta = 0.5$mm and two different strain rates. Black color FTACLES model, blue color directly filtered flamelet.

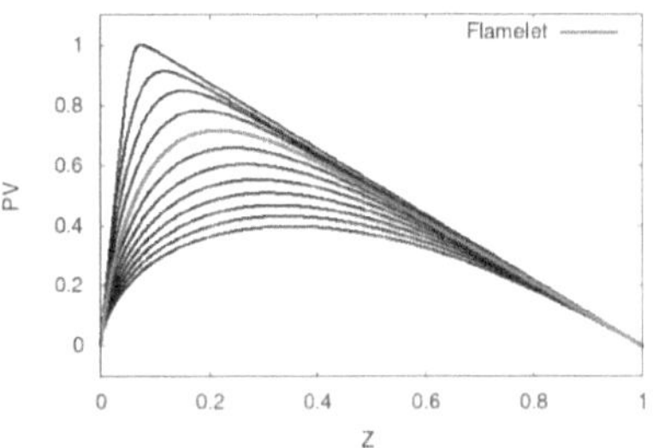

Figure 3.32: Flamelet identification within a manifold

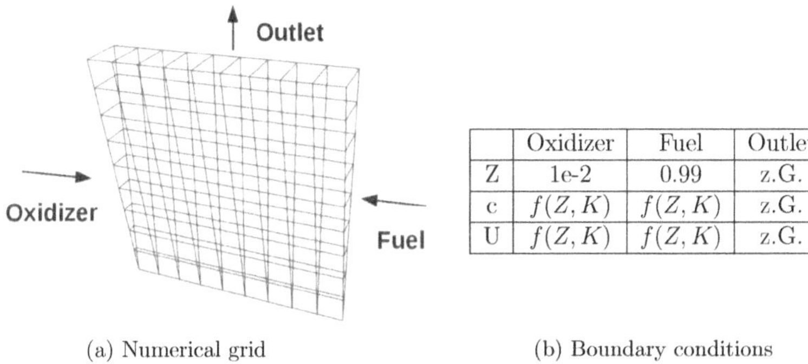

	Oxidizer	Fuel	Outlet
Z	1e-2	0.99	z.G.
c	$f(Z, K)$	$f(Z, K)$	z.G.
U	$f(Z, K)$	$f(Z, K)$	z.G.

(a) Numerical grid   (b) Boundary conditions

Figure 3.33: Two dimensional case. z.G. stands for zero-gradient.

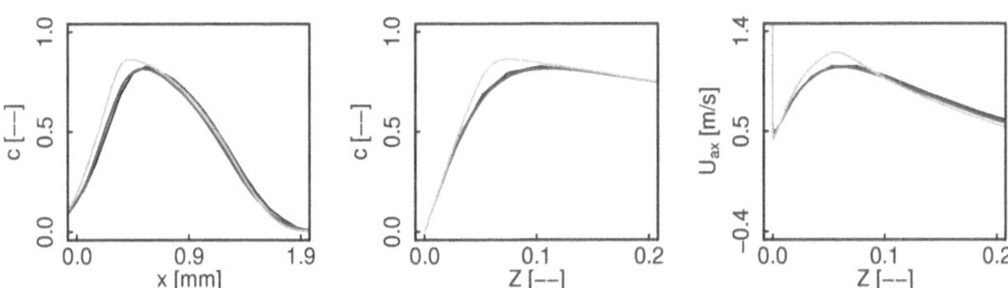

Figure 3.34: Filtered progress variable and velocity profiles in physical and Z space, with $\Delta = 0.5$mm and $K = 600s^{-1}$. Black color FTACLES model, blue color directly filtered flamelet, green color unfiltered flamelet.

The filtering effect increases with the profile thickness, so that OH decreases to one half of its value and ω_c to one third. The lower K close to $Z_{st} = 0.055$ results into a slight under estimation of the source term, and the inverted effect for OH. Far away from the reaction zone the deviation in K prediction has a negligible impact on the scalar prediction as both profiles tend to zero at the extremes.

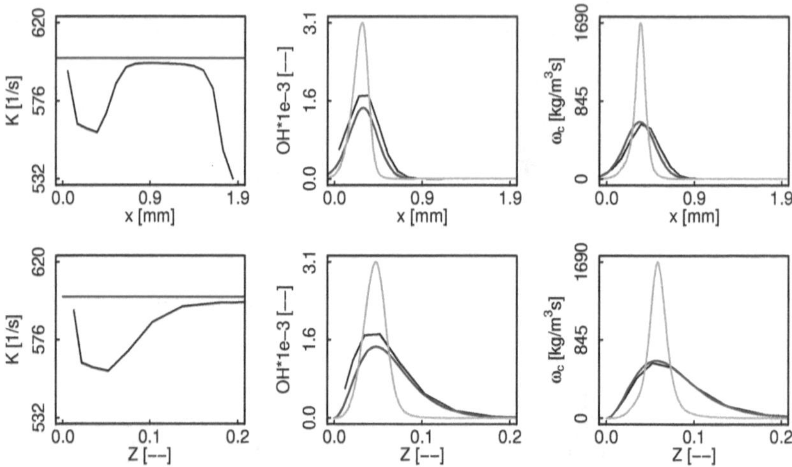

Figure 3.35: Retrieved strain rate, OH and progress variable chemical source term for $\Delta = 0.5$mm and $K = 600s^{-1}$ in physical and Z space. Black color FTACLES model, blue color directly filtered flamelet, green color unfiltered flamelet.

3.6 Conclusions

This chapter addressed the filtering of a non-premixed flamelet. First the problem mathematical formulation was introduced. Subsequently the effects of the filtering operation on a non-premixed flamelet were appraised. The flame structure modification was described considering an individual flamelet and the entire manifold. The following distinctive features were identified:

- The filter effect varies considerably along the manifold as function of K. This results from the flame thickness reduction at higher strain rates.

- The filter effect also differs for each variable within a flamelet, depending on the profile shape and thickness. As a result, there might be a non linear relation between Z, c and the retrieved species φ under the filtering operation. For big thickness difference and above a given filter size, this will significantly alter the variable parameterization and so the species predictions.

- Non-monotonic variables are substantially modified due to the trimming of the peak value. Considering the progress variable, this transformation shifts the flamelets downwards within the manifold. Therefore a fixed point $P_0 = (Z_0, c_0)$ will be described by a different flamelet depending on the filter size employed.

- The deviation in the retrieved variables prediction increases with the filter size. Two aspects dictate the magnitude of the discrepancy: first the significance of the flamelets deviation within the manifold, and second the considered species filter sensitivity.

- Depending on the filter size, a portion of highly strained flamelets replaces part of the unsteady solutions, fulfilling however the same continuation function towards a pure mixing condition.

The filtering effect over a premixed and a non-premixed manifold was then addressed. First the premixed case was introduced and distinctive features were identified. Subsequently premixed flamelets were compared against the corresponding composition iso-Z contours extracted from a non premixed manifold. The source term undergoes a higher peak reduction with increasing filter size for the non-premixed case. Considering non-premixed filtering as a continuation method, the comparison against premixed flamelets indicates that care should be taken when non-premixed filtered flamelets are shifted to lower manifold regions. The big filter effect results into rather homogeneous counterflow profiles which might significantly affect the retrieved variables predictions, e.g. the species or thermophysical properties.

The comparison of species contours further enlightened the different filter effect for the two regimes. The relation between $Z - c$ is altered on non-premixed flames, thus the species behavior in chemical space is inevitably modified with increasing filter size. For the premixed condition on the contrary, c transformation impacts exclusively the physical space. It follows that the flame structure transformation in chemical space depends on the correlation between the considered species and the progress variable. For highly correlated variables, e.g. T, the filtered manifold strongly resembles the unfiltered one independent of the filter size, hence decoupling the species prediction from the flame thickness. This is a fundamental issue as it completely modifies the interpretation of a filtered flame on the two approaches.

The manifold transformation under a presumed β-PDF was compared against the non-premixed filtering. The main differences were highlighted, i.e. the operation takes place in mixture fraction space, and the profile modification is based on the mixture fraction fluctuations, thus being completely independent from the strain rate. The analysis shows the existence of corresponding Z_v^2-Δ, pairs where the flamelet transformations appear to be considerably similar for different considered species. The results suggest the possibility of interpreting the filtered profiles in an analogous relation with the PDF integration, where the equivalent SGS fluctuations directly increases with the strain rate for any given filter size.

Chapter 4

Laminar diffusion flame

When simulating turbulent diffusive combustion either in RANS or LES context, closure strategies are necessary to deal with unresolved terms. The estimation of the chemical source term becomes a particularly challenging issue, as the flame front cannot be resolved on the numerical grid, and the complexity augments further due to the interaction with turbulence structures. Thus, the assessment of a laminar flame configuration delivers valuable insight on the ability of a given approach to correctly describe the physics, decoupling it from SGS modeling.

The goal of this chapter is to elucidate the physical interpretation of the filtering operation by appraising the performance of the filtered tabulated chemistry model over a coflow laminar diffusion flame. Recalling that the ultimate objective of the model is to be employed in the LES context, this study is a necessary intermediate step to substantiate the consistence of the formulation. Satisfactory results in this stage would then justify the further consideration of an efficiency function taking into account the modified interaction between the filtered flame front and the subgrid turbulent structures.

This chapter demonstrates the soundness of this thickening flame front approach on a 3-D real laminar non-premixed coflow flame. The model results are compared against the direct filtering of the fully resolved laminar diffusion flame showing that the formalism adequately describes the underlying physics. The study reveals the importance of the one-dimensional counterflow flamelet hypothesis, so that the model activation under this condition is ensured by means of a flame sensor. The consistent coupling between the model and the flame sensor adequately retrieves the flame lift-off and satisfactorily predicts the profile extension due to the filtering process. The chapter is structured as follows:

- Section 4.1 introduces the study case consisting on a laminar diffusion coflow flame. First non-premixed filtered tabulated chemistry results are presented and the reasons for the observed deviations are discussed.

- Section 4.2 proposes a model activation strategy employing a sensor S based on the flame dimensionality.

- The model's performance is assessed and the results are then discussed in Section 4.3, where the fundamental role S plays on the non-premixed filtered tabulated chemistry model is highlighted.

- Concluding remarks are presented in Section 4.5.

4.1 Flame characterization

A laminar coflow methane-air diffusion flame numerically analyzed in [144] has been simulated to assess the model performance in absence of turbulence. The burner configuration consists of a central fuel jet with a mixture composition of 55%CH_4 and 45%N_2 in molar fractions, surrounded by an air coflow, as shown in figure 4.1a. The fuel velocity is defined by a parabolic profile, where $u_{max} = 0.23$ m/s, while for the coflow air a plug flow with a constant velocity equal to the maximum fuel velocity is employed. The jet's inner radius is $r_{Fuel} = 6$ mm, the outer coaxial tube has a radius of $r_{Air} = 27.5$ mm, and the domain length is 90 mm. These conditions lead to a lifted flame as shown in Figure 4.1b which reproduces the finding in [144].

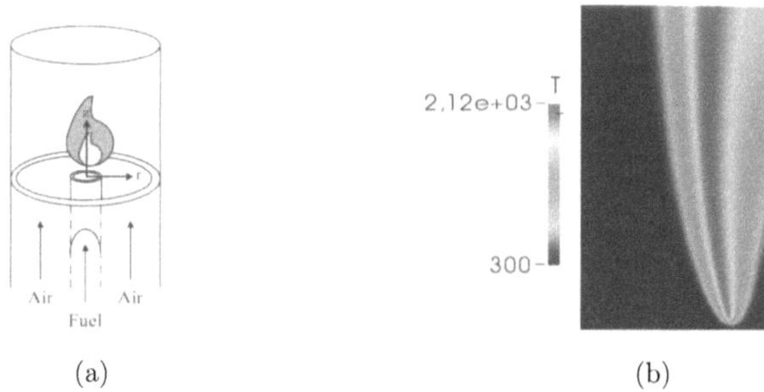

(a)
(b)

Figure 4.1: Burner configuration [144] and temperature field

The numerical grid is two-dimensional, axisymmetric, structured, constantly spaced along the axial direction, while radially a uniform spacing is employed for $r \leq r_{Fuel}$ and outwardly increasing cell size is used for $r_{Fuel} < r < r_{Air}$. Three different refinement levels are used. $NU1$ is a finer mesh than the one used in [144], which proved to have an appropriate resolution. Systematic coarsening, with an increasing factor $F \approx 2$ in each direction is applied to obtain $NU2$ and $NU3$. The grid details are indicated on table 4.1, where the last column corresponds to the average cell size at the stoichiometric condition throughout the domain, defined as $h_{st} = \langle h | Z = Z_{st} \pm 0.01 \rangle$.

Table 4.1: Numerical grid parameters.

Mesh	NoCells	h_{min} [mm]	h_{max} [mm]	h_{st} [mm]
NU1	127658	0.1	0.23	0.13
NU2	33300	0.2	0.45	0.24
NU3	9225	0.4	0.82	0.45

The case is first solved employing the detailed chemistry solver *laminarSimpleSMOKE* [34],[35],[36] with unitary Lewis number assumption. The scatter of c as a function of Z (computed according to the modification proposed by [9]) is shown in Figure 4.2. In steady flamelets any single trajectory connecting the fuel and the oxidizer ($Z = 0$ and $Z = 1$) can exclusively describe either an ignited or a pure mixing condition. The last ignited flamelet is here shown in red and unsteady flamelets are employed to describe the points lying below this threshold, which correspond to the flame lift-off, appreciated on the temperature profile shown in Figure 4.1b.

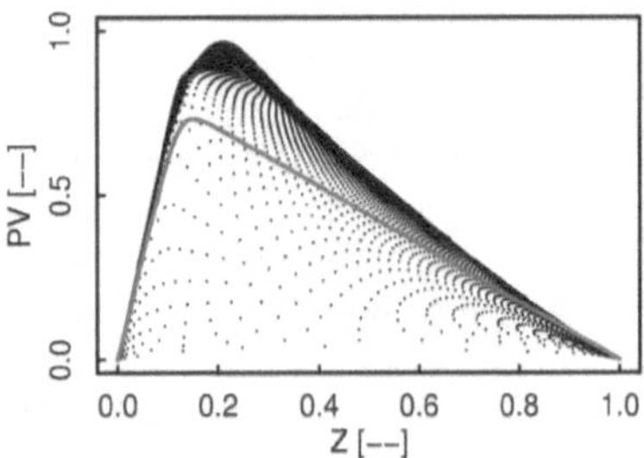

Figure 4.2: Progress variable scatter plot in mixture fraction space. Red line corresponds to the extinction flamelet, $K = 482s^{-1}$.

The manifold is generated using the FGM method with non-premixed flamelets. The strain rate has been varied from $0.03s^{-1}$, which can be considered as chemical equilibrium, up to $482s^{-1}$, after which extinction takes place. In order to fully span the composition space, time-dependent flamelet solutions have been added until the pure mixing line is reached. The flamelets have been generated employing the detailed chemistry solver CHEM1D[136], where the $GRI - 3.0$[134] detailed chemical mechanism has been considered with $Le = 1$ assumption. The c species combination proposed in [144] has been adopted, though highly promising techniques such as PCA (Principal Component Analysis)[108] might as well offer appealing alternatives to be appraised in future studies. The progress variable is defined as $c = \sum \alpha_i^* Y_i / \sum \alpha_i^* Y_{i,eq}$, where the weight factor $\alpha_i^* = 1/MW_i$ is the inverse of the molecular weight, and the selected species i are CO_2, H_2O and H_2.

4.1.1 Filtered tabulated chemistry problem

The case was solved applying the non-premixed filtered tabulated chemistry formulation. This procedure was carried out for three different filter sizes: $\Delta = 2,3,4$ mm. For all of the cases an unphysical mixture fraction evolution, which worsened with increasing filter size, was obtained. The results corresponding to $\Delta = 2$mm are shown in Figure 4.3 where an already evident deviation from the reference can be seen, while not all the trends of the original solution have been lost.

Since the reason for Z distortion relies on the model correction terms, which alter its passive scalar nature, their behavior is shown in Figure 4.3b. Initially both of the terms are negative and reach their highest magnitude close to the inlet, resulting in a steep Z decay. The subsequent sign change of the diffusive term is responsible for the inflection point on the Z profile starting from the first dashed line. The summation of the correction terms becomes positive, causing an unphysical Z increase, and then it balances. No trend difference with respect to the reference solution can be appreciated on the Z profile in the vicinity of the second dashed line, which indicates the point where the correction terms finally vanish.

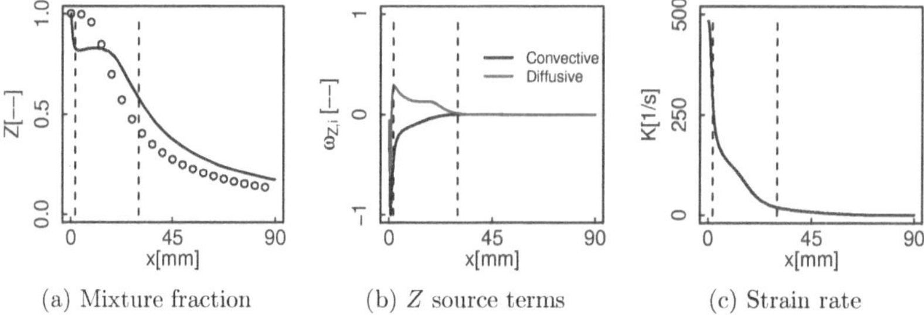

(a) Mixture fraction	(b) Z source terms	(c) Strain rate

Figure 4.3: Centerline profiles of mixture fraction, normalized model correction terms and strain rate for $\Delta = 2$mm. Solid lines represent the FTACLES results, circles correspond to the reference solution.

The strain rate K at the oxidizer side is read for each flamelet and it is tabulated as $K = f(Z, c)$. This value is not modified during the filtering operation and so it serves as a reference to the flamelet from which the filtered profile has been obtained. This is a quite convenient feature, since in this way the filter effect can be individuated and useful information can then be extracted for instance by tracking the displacement of a given K. The simultaneous consideration of figures 4.3b and 4.3c delivers valuable insight concerning the difference in the model behavior along the center line. The variables appear to be correlated, as high correction values coincide with high strain rates. This observation points out to a thorough consideration of the diffusion flamelet manifold behavior, since the filter effect considerably changes among the flamelets due to the wide range of flame thicknesses.

Figure 4.4 shows the profile transformation undergone by the two extreme steady flamelets of the current manifold, i.e. the equilibrium and the last ignited flamelets, under three different filter sizes. The equilibrium flamelet has a large thickness and it is essentially not affected by the filtering operation so that the three profiles overlap becoming indistinguishable. This behavior categorically changes for $K = 482s^{-1}$, that completely changes its shape for each filter size. For instance using $\Delta = 5$mm c profile loses its peak close to Z_{st} and is symmetrically distributed over the domain.

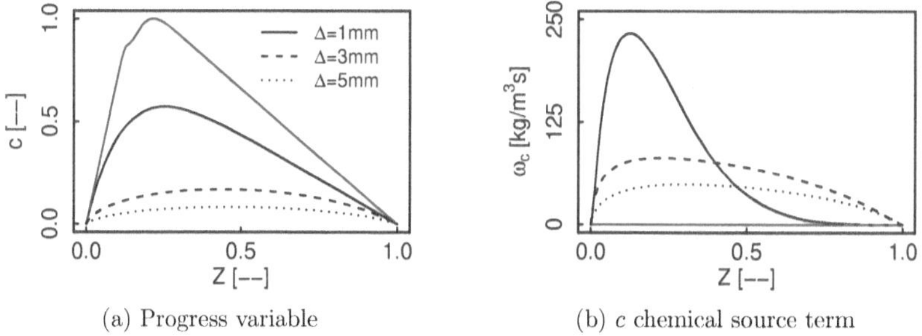

(a) Progress variable	(b) c chemical source term

Figure 4.4: Flamelet transformation for two different strain rates and three filter sizes. Blue lines $K = 0.03s^{-1}$ and black lines $K = 482s^{-1}$. The three blue lines superimpose so that only one profile can be perceived.

The source term increases with K, as the reaction must proceed faster due to the diminished residence time, so that for these extreme flamelets $\dot{\omega}_c$ does not even share the order of magnitude. Moreover the same phenomenon described for the progress variable is observed on the source term at $K = 482s^{-1}$, from where the reactive zone can no longer be recognized as $\dot{\omega}_c$ extends all over the domain. Deviations between the approximated and resolved diffusive and convective transport terms arise for the last ignited flamelet, while the equilibrium flamelet is not modified by the filtering operation, and consequently the correction terms are zero. This feature, namely the high filter sensitivity of the last ignited flamelet opposed to the unaltered equilibrium flamelet behavior explains the disappearance of the correction terms after the second dashed line in Figure 4.3b, as K goes beneath a low threshold value.

When assembling a manifold of counterflow flamelets c decreases with K for any given Z, except for the extremes where all the profiles converge. On a real coflow flame the fuel side might considerably distinguish from the counterflow setup due to geometrical constraints and flow orientation. In such a situation, c evolution in the actual flame describes a K profile that departs from a high value at the fuel side and that subsequently decreases towards c_{max} location, where the flame structure resembles the counterflow configuration. This is the case at the centerline, so the obtained profile is coherent with the manifold intrinsic features, and the observed Z deviations obtained with the filtered tabulated chemistry approach result from the substantial role that the strain rate parameter occupies in the model performance.

These outcomes, namely the flamelet strain rate filter sensitivity and the manifold-conditioned centerline strain rate profile do not only elucidate the results exhibited in Figure 4.3, but moreover point out the need of an additional tool to determine the use of the model correction terms. Some characteristic features of a coflow diffusion flame will then be analyzed aiming to better comprehend the underlying reasons why the intrinsic assumptions of the filtered tabulated chemistry model might not be fulfilled.

4.1.2   Flame multidimensionality

There are phenomena taking place on multidimensional non premixed laminar flames which do not coincide with the fundamental flamelet assumptions, as has already been pointed out by several authors[131]. The one dimensional hypothesis correctly describes an idealized planar counterflow flame, but for a coflow flame the consideration of multidimensional effects results imperative. For instance, it has been shown that under given simplifications, e.g. radiation and differential diffusion being neglected, steady flamelets adequately recover the main flame features, while discrepancies are observed principally along the centerline [25]. An exhaustive comparison between detailed chemistry and FGM results for the current study case is presented in [144]. The authors show that when the assumption $Le = 1$ is employed, even though FGM and detailed chemistry results coincide, tangential diffusion and curvature effects play a significant role at the center line. Their repercussion becomes detectable as soon as the diffusion coefficient is distinctively calculated for each species, and FGM results deviate from the detailed chemistry ones.

Aiming to decouple the multidimensional effects from any flame front assumption, the results presented in this section correspond to the solution employing the detailed chemistry solver. The one-dimensional flamelet hypothesis has then been assessed by means of the mixture fraction and species gradients. Considering that the current case corresponds to a coflow flame injected in the x direction, the y coordinate can be said to coincide with the mixture fraction isolines. Defining the gradient angle as $\theta_{\nabla\varphi} = tan^{-1}(\nabla\varphi_x/\nabla\varphi_y)$, $\theta_{\nabla\varphi} = 0$ would correspond to the one dimensional counterflow flame hypothesis, while $\theta_{\nabla\varphi} = \pm 90$ describes a gradient evolving along the axial direction. Figure 4.5 shows the computed angle along three different radial sections.

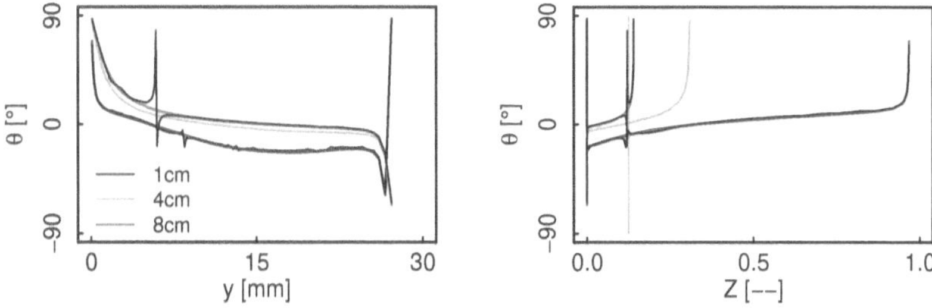

Figure 4.5: Gradient angle profiles in physical and mixture fraction space at different radial sections. Colored lines correspond to ∇Z , black lines are ∇CO_2.

The radial gradient dominates and the angle tends to $0°$ in the majority of the domain. The discontinuity in ∇CO_2 corresponds to the inflection point in CO_2 profile close to $Z = Z_{st}$. Due to the monotonic evolution of the mixture fraction, this situation does not occur for ∇Z and the counterflow hypothesis still holds in this region of maximum chemical activity. Nevertheless, a deviation from this behavior is obtained at the extremes, i.e. close to the centerline and at the side, where $\theta_{\nabla Z} \approx 90°$.

Keeping in mind the results shown in section 4.1.1, the attention will then be focused on the centerline, whose behavior can be appreciated in more detail in Figure 4.6. The dashed line delimits a zero gradient region close to the inlet, so the angle estimated before this point should not be taken into account, due to lack of physical meaning. As the gradients starts to grow the radial term initially plays a non-negligible role that almost evens up with the axial component. This y-oriented contribution significantly decreases downstream so that the flame front becomes basically perpendicular to the flow.

Being aware of the difference between both simulations, some insightful remarks can be done in relation with the filtered tabulated chemistry results: the zone of the steep Z decay in Figure 4.3a delimited by the first dotted line lies within the zero gradient region in the detailed chemistry. Like this, the addition of correction terms hinders the initial constant value transport of Z by means of an erroneous model activation. The second aspect to observe is that the vanishing of correction terms, delimited by the second dotted line, coincides with the decrease of the Z gradients. It is then useful to emphasize that the variable's higher variation zone as well as higher model activity never fulfill the one-dimensional hypothesis.

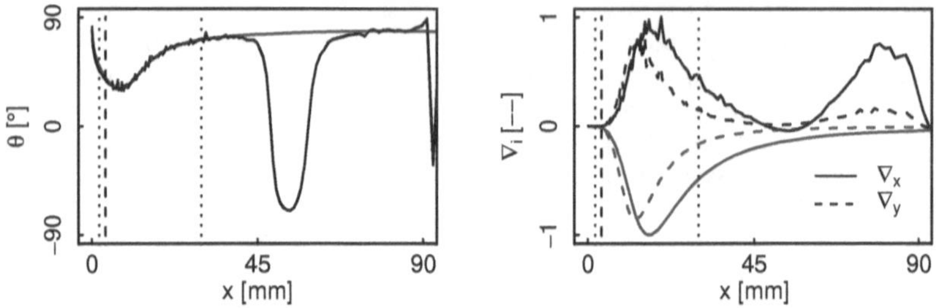

Figure 4.6: Centerline gradient angle and normalized gradient components. Blue lines Z, black lines CO_2, dashed line zero gradient limit, dotted lines correspond to the zones indicated Figure 4.3.

4.1.3   Flame strain and unsteadyness

The scalar dissipation rate χ has been computed using equation (2.80), and both its normalized value as well as the normalized χ_0 from equation (2.82) are shown in figures 4.7b and 4.7a. The scalar dissipation rate equals zero before the first dashed line, the jet core vanishes and χ starts to increase until it reaches a peak, after which it decreases once again. Recalling from Figure 2.3 the assumed χ shape along a diffusion flamelet reaches its highest value at $Z = 0.5$, in agreement with the function $F(Z)$, while in this case on the contrary, χ_{max} is located at $Z = 0.7$.

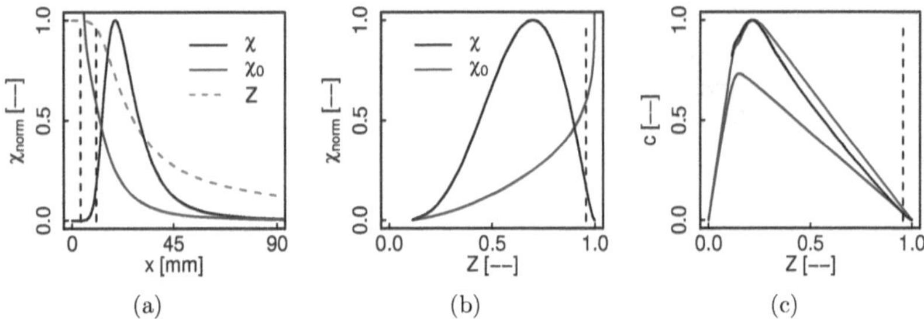

Figure 4.7: Detailed chemistry centerline profiles in physical and mixture fraction space.

The χ_0 profile is characterized by an asymptotic behavior at the inlet, followed by a decreasing and smoothened evolution downstream. The role of χ_0 is analogous to K, i.e. it serves as a flamelet tag, hence the importance of this result relies on the fact that χ_0 shows the same trend as that observed in Figure 4.3c. This is a fundamental result, since it proves that given the centerline evolution of χ in this flame, if a counterflow flamelet approach is used to describe it, the highly strained region of the manifold will be always accessed when solving the lifted and ignition region. A corollary being that the filtered tabulated chemistry method should then be able to identify and appropriately handle this transition.

4.2 Flame sensor

4.2.1 Justification

A flame sensor can be employed as a tool to identify the region of maximum flame chemical activity. For the case of combustion modeling one of the applications of such a sensor is to turn on or off the model by distinguishing between reacting and pure mixing (non reacting) zones. In the ATF context for example, flame sensor definitions based either on c[44] or on its source term Ω_c[130] have been used. The thickening factor is then multiplied by the sensor, so that the latter being equal to zero means that there is no need to apply any correction. In such a scenario the variable transport is fully determined by convection and diffusion which can be adequately resolved on the numerical grid.

The one-dimensional counterflow configuration assumption is a key issue in the derivation of the non-premixed filtered tabulated chemistry model. This constraint does not rely on the filtering principle itself, which can be applied to any arbitrary configuration, but it is due to the fact that the model correction terms have been determined as to satisfy the equations describing this specific configuration. The results presented in section 4.1.1 showed big model deviations at the centerline, precisely there where section 4.1.2 highlighted the presence of multidimensional phenomena, which is indeed reasonable given that the model falls out of scope. In this way, an approach is proposed which assesses the model performance by activating the correction terms exclusively in the counterflow regions, taking advantage of a sensor based on ∇Z.

This formulation has been chosen exploiting the fact that the flamelet thickness decreases with the strain rate, consequently modifying the gradients. Based on one-dimensional flamelet characterizing features two assumptions are made, whose adequacy will be appraised afterwards. Defining a point P described by the pair (Z_P, c_P), the first hypothesis states the uniqueness of the mixture fraction gradient, i.e. non-coincidence for different configurations. Second, it is assumed that the maximum possible gradient will be obtained on a counterflow condition. In other words, if P does not comply with a counterflow condition, the flame front will unquestionably possess a lower strain rate than the counterflow flamelet corresponding with these coordinates.

4.2.2 Proposed definition

Considering that on a flamelet-based approach the flame front can be visualized as a set of thin mixture fraction iso-surfaces, it follows that for every single point, the gradient of the transported mixture fraction ∇Z_{sim} should coincide with that of the accessed flamelet ∇Z_{fl}. The underlying assumption being that both the flamelet and the real flame front share the same resolution, which in general is not the case, and as a matter of fact is the motivation for the FTACLES model.

The detailed chemistry solution has been computed employing a regridding strategy in order to adequate the mesh in the regions of greater gradients. Before performing the filtering operation this flamelet is reinterpolated over a new mesh, each cell size being the minimum between the original cell size and $\Delta/5$. This is necessary in order to adequately compute the correction terms and for the reconstruction of the retrieved variables $\varphi = f(Z, c)$. However this has deep consequences in ∇Z

computation when compared to a lower resolution mesh. To illustrate this, Figure 4.8 compares the normalized mixture fraction gradient behavior of two strained flamelets when varying the numerical grid resolution. First each flamelet was directly filtered with $\Delta = 2\text{mm}$ and ∇Z was computed. Three different Δ/h ratios were defined and constantly spaced progressions mimicking coarser numerical grids were generated. The filtered flamelet Z profile was projected over these new domains and ∇Z were subsequently computed employing finite differences.

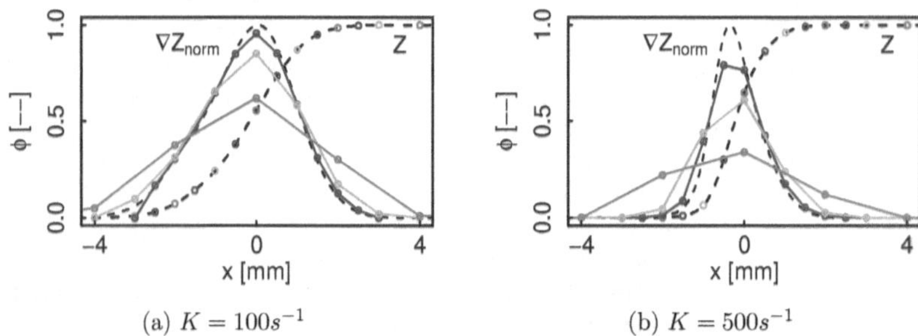

(a) $K = 100s^{-1}$ (b) $K = 500s^{-1}$

Figure 4.8: Numerical grid resolution effect over the mixture fraction gradient for two $CH_4\text{-}Air$ flamelets with $\Delta = 2\text{mm}$. Blue line $\Delta/h = 4$, green line $\Delta/h = 2$, red line $\Delta/h = 1$, dashed lines are the reference filtered flamelet.

Given that the domain is described with less points, as the ratio Δ/h decreases so does the precision in the reactive zone description. This effect is much stronger as K increases, which is indeed coherent with Figure 3.2 that depicts the wide thickness range along a non-premixed manifold. For instance $\Delta/h = 2$ reproduces quite well ∇Z profile for $K = 100s^{-1}$, while a considerable deviation is observed for $K = 500s^{-1}$, and this discrepancy will vary depending on the specific point around which the gradient is computed. Based on these considerations a flame sensor S comparing the simulation and reference flamelet mixture fraction gradients is proposed as

$$S = H(\nabla Z_{sim}/\nabla Z_{fl} - ct) = \begin{cases} 0 & \text{if} \quad \nabla Z_{sim}/\nabla Z_{fl} < ct \\ 1 & \text{if} \quad \nabla Z_{sim}/\nabla Z_{fl} \geq ct \end{cases}, \tag{4.1}$$

where H is the Heaviside or unit step function and ct corresponds to a tolerance that takes into account the varying effect of numerical grid resolution along K. A mixture fraction offset σ_Z is defined in order to avoid dividing by $\nabla Z_{fl} = 0$ at the extremes, i.e. $Z = 0$ and $Z = 1$. In this way the sensor is active within the range $Z \in [\sigma_Z, 1 - \sigma_Z]$, and $\sigma_Z = 10^{-4}$ has been used, in agreement with the table resolution towards the edges.

An alternative approach would be to impose the number of desired points on the flame front, so stating a ratio between the filtered flame thickness and the maximum filter size. Two possibilities exist afterwards: if the table remains two-dimensional this will constrain Δ to the highest strained flamelet and this might deliver not practical cell size values; while the second alternative is to construct a three-dimensional table, with the corresponding additional numerical complexity. Given the inconveniences both of the options present, it has then been decided to proceed with the proposed sensor. It must however be considered that under the current definition a limit filter size might exist for every flamelet, above which the resolution might be too low as to correctly characterize the gradient.

4.2.3   Continuation flamelet identifier

This gradient-based model activation aims to evaluate the validity of the counterflow assumption, and so it addresses the first discrepancy emerging from the previously performed analysis. The second identified issue consisted on the characterization of the flame lift-off zone, described by unsteady flamelets on an FGM approach, which challenges the ability of the non-premixed filtered tabulated chemistry to deal with this phenomenon. Given that it is not straightforward to theoretically sustain the sensor adequacy under this condition, an interpretation rather than a justification of how the sensor might perform is presented. Thus Figure 4.9 recalls the case already analyzed in Section 3.2.5, namely two filtered steady flamelets whose profiles lie over unfiltered unsteady profiles.

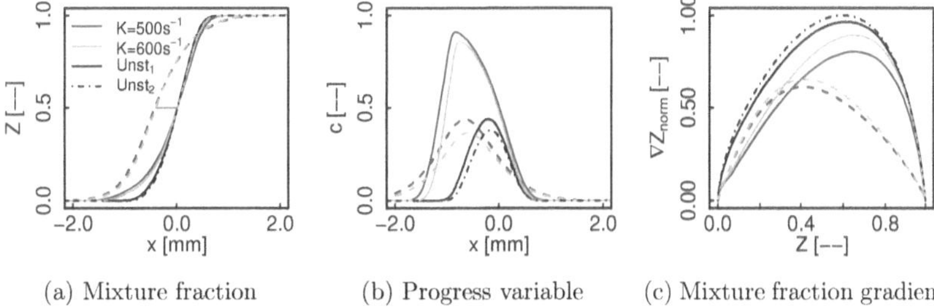

(a) Mixture fraction   (b) Progress variable   (c) Mixture fraction gradient

Figure 4.9: Profile distinction for Z space analogous trajectories of steady filtered and unsteady unfiltered flamelets when observed in the physical space. Solid lines unfiltered profiles, dashed lines filtered profiles, and the red line is a separation included to better individualize the curves.

The first aspect that calls the attention is that even though the variables behavior shown in Figure 3.18 shows a considerably good accordance between the unsteady and the filtered steady flamelets projected in Z space, their physical space profiles significantly differ as shown in Figure 4.9a. This can be understood considering that the unsteady flamelets will tend to keep the same domain length as their originating flamelet, namely the extinction one, while the filtered ones will expand respect to their unfiltered counterpart. As a result of this for the same $Z - c$ combination the ∇Z profiles are considerably lower for the filtered flamelets than for the unsteady ones. The corollary is that in order for the filtered steady flamelets to perform a continuation role, ∇Z_{sim} should not only be smaller than that of the reference flamelet, but it should as well be significantly smaller than the original unsteady counterpart. Otherwise, the point describes a steady filtered flamelet, with the corresponding model correction terms.

Expanding on this aspect, the sensor can be more globally interpreted as a flamelet identifier. In this way, independent of the location in Z space, an active sensor describes a point which behaves as a counterflow filtered flamelet, while the contrary corresponds to a point that obeys the variable transport while not fulfilling the model hypothesis. For instance, another condition under which the flame sensor results useful, and that has not been discussed so far are the constant mixture fraction regions. This might happen at the domain extremes, e.g. at the fuel core, but might as well take place inside the Z domain, for a multi-streamed configuration. In any of the cases the constant valued Z zone imposes a particular constraint within the system, so that the addition of the model correction terms would deliver unphysical results. The sensor appropriately handles this situation, since ∇Z tends to zero, and consequently the model deactivation is assured.

4.3 Filtered tabulated chemistry results

The reference solution was obtained following the procedure suggested in [144] employing a FGM manifold. The case was then solved applying the novel flame sensor with $ct = 0.9$ within the non-premixed filtered tabulated chemistry formulation. The equations presented in Section 3.1 have been written considering the LES framework, so that the filtered terms correspond to the resolved part. For the current laminar case this transformation, and the use of filtered tables is equivalent to perform a filtering operation over the whole domain. Exploiting this quite convenient feature, the results could be compared against the ones obtained applying a Gaussian filter to the reference solution. This procedure was carried out for three different filter sizes: $\Delta = 2,3,4$ mm.

4.3.1 Model performance

For each filter size, simulations were carried out using the three numerical grids. The results appear to be mesh independent. The minimum amount of grid points relative to the filter size corresponds to $\Delta/h_{st} \approx 4.5$, which adequately resolved the gradients. The results match correctly the reference data as can be observed in Figure 4.10.

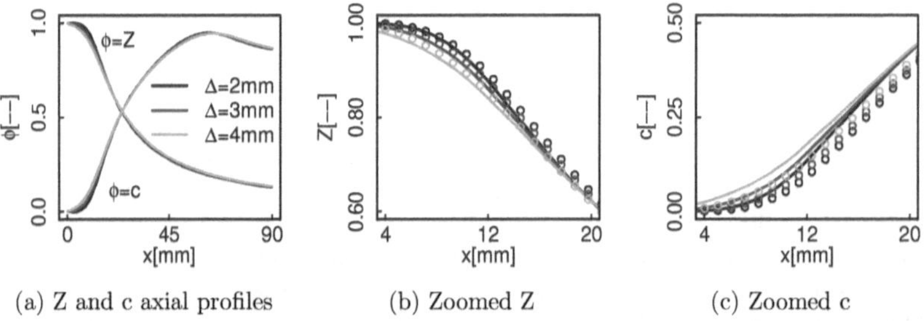

(a) Z and c axial profiles   (b) Zoomed Z   (c) Zoomed c

Figure 4.10: Filtered axial profiles of mixture fraction and progress variable (a); zoom of region with bigger filter effect for Z (b) and for c (c). Solid lines represent the filtered tabulated chemistry results, circles correspond to the filtered reference solution.

The ignition zone, which corresponds to the region with the biggest gradients along the centerline has been zoomed in figures 4.10b and 4.10c. It can be observed

that the model is able to adequately retrieve the profile transformation resulting
from the filtering process. The mixture fraction decay is overpredicted with a maximum relative error $RE_Z = 5.1\%$, corresponding to $\Delta = 4$mm. From the physical
point of view this variation entrains a decrease in the unignited flame core which is
shifted backwards and induces the displacement of c profile.

The filtering process decreases the peak values for non-monotonically evolving
functions. The right column of Figure 4.11 shows OH profiles correctly matching
the filtered reference solution. The peak OH value is predicted with a maximum
$RE_{OH} = 6.7\%$, which occurs at $x = 8$cm and $\Delta = 4$mm. The progress variable
non-monotonic profile shows lower sensitivity to the filtering process as a result of
its higher thickness. Following the small mixture fraction gradients in the radial
direction, modest differences, not surpassing an absolute error of $AE_Z = 0.02$ are
found at $x = 1$cm, while downstream the profile change is almost negligible as shown
in the left column of Figure 4.11.

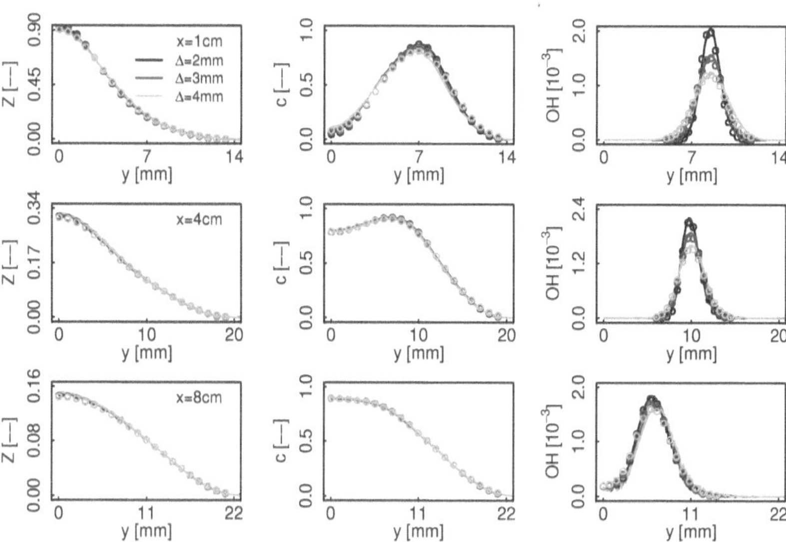

Figure 4.11: Filtered radial profiles. Solid lines represent the filtered tabulated
chemistry results, circles correspond to the filtered reference solution.

Recalling that K is saved in the table as a tag to the original flamelets, it provides
information on the manifold region being accessed. In this way, the almost absent
filter effect over c for $x = 4$cm and $x = 8$cm can be explained from Figure 4.12
due to the low K, which as a matter of fact decreases downstream. The flame front
becomes slower, possesses a bigger thickness and consequently is less sensitive to the
filtering operation. The filter effect becomes then only perceptible through c profile
deviations at the rich side at $x = 1$cm.

Focusing on the zone where c reaches its peak, it can be acknowledged that
K profiles converge to almost the same values. It is known that one-dimensional
flamelets in general terms can adequately describe a diffusion flame, and that dif-

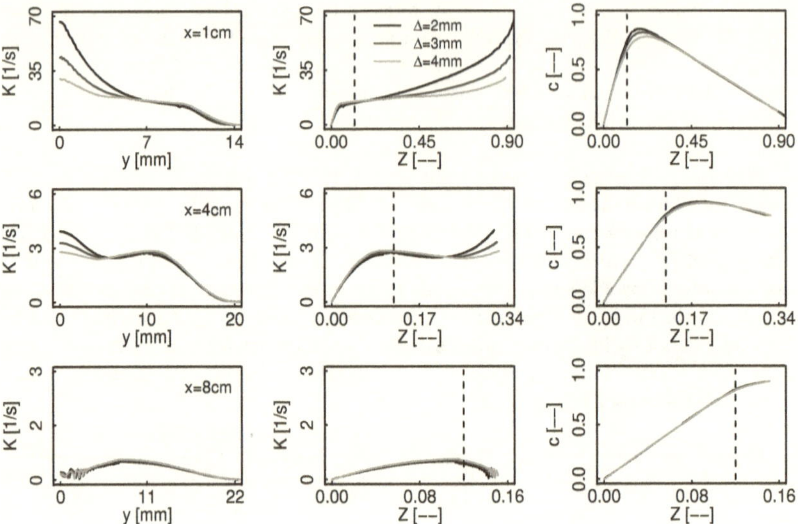

Figure 4.12: Strain rate and progress variable radial profiles in physical and mixture fraction space. Solid lines radial profiles, vertical dashed line Z_{st}.

ficulties emerge close to the centerline due to mutidimensional effects. However these results suggest that far from the centerline not only the reactive zone can be described as a counterflow flamelet, but moreover, that the profile variation and smoothening in the most sensitive region can be understood as the filtering of the same $Z - c$ trajectory.

In order to evaluate the role of the model itself, the case was resolved employing filtered tables but deactivating the correction terms. Figure 4.13a shows that along the centerline the use of different filter sizes without additional correction terms does not exert any change neither on Z nor on c profiles, and they converge to the original unfiltered values. The same behavior is appreciated as well for radial Z profiles, which remain unaltered in Figure 4.13b. These results outline the model correction terms relevance in order to retrieve the correct profile expansion. Figure 4.13c shows that the decrease in peak values, on the contrary, is mostly dominated by the filtering operation and not by the addition of correction terms. This can occur in two ways: for a transported variable, e.g., c, this is the effect of a lessened $\dot{\omega}_c$ as a result of the filtering operation; while for the tabulated values, i.e. $\varphi = f(Z, c)$, the profile modification will take place as an simultaneous effect of decreased c and of the modified φ profiles themselves.

4.3.2 Sensor response

The sensor tolerance is bounded between two conditions: $ct_{Max} \leq ct \leq 1$ where S is permanently deactivated, and $0 < ct \leq ct_{Min}$ where it is always active. In order to assure the activation flexibility and simultaneously guarantee the correctness of the results, a sensitivity analysis was carried out, and the optimal ct value was defined

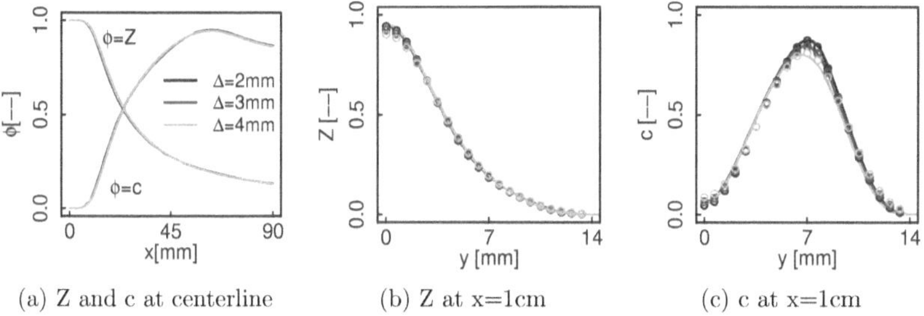

(a) Z and c at centerline   (b) Z at x=1cm   (c) c at x=1cm

Figure 4.13: Mixture fraction and progress variable centerline and radial profiles for three different filter sizes. The manifolds are generated by direct flamelet filtering, the filtered tabulated chemistry correction terms being omitted.

as the minimum for which two successive ct curves converged. Figure 4.14 depicts the model response for the three different filter sizes, from where the tolerance was determined as $ct = 0.9$. The model sensitivity increases with the filter size, thus while $ct = 0.9$ and $ct = 0.8$ profiles converge for $\Delta = 2$mm, a slight deviation is observed for $\Delta = 3$mm and they are clearly distinguishable for $\Delta = 4$mm. In order to further illustrate ct relevance, the three dashed lines evidence the increasing model distortion that results from decreasing tolerance values.

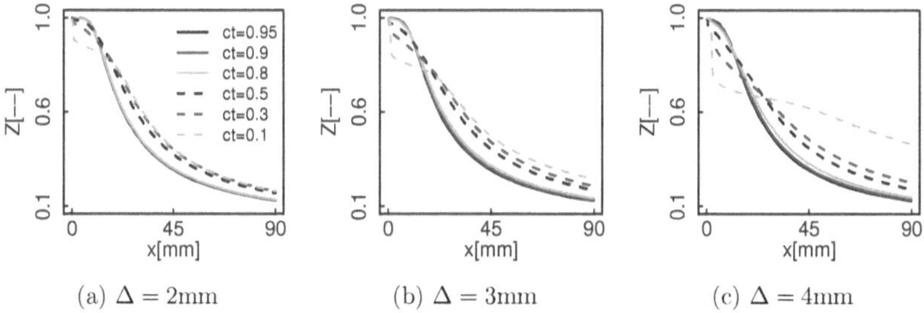

(a) $\Delta = 2$mm   (b) $\Delta = 3$mm   (c) $\Delta = 4$mm

Figure 4.14: Effect of the sensor ct variation, assessed on the Z centerline profiles for three different filter sizes.

As to gain further insight on the model's behavior throughout the flame, Figure 4.15 presents the correction terms behavior for $\Delta = 4$mm along two different radial sections. This filter size is chosen in order to better appreciate the correction terms, but the same trends have been observed for the other two smaller filter sizes. The first aspect to recall is the reduction in the source terms, which obeys two different reasons both linked to the downstream strain rate decrease along the flame front. On one side, the chemical source term decreases directly with K, which can be interpreted as the reactions becoming slower; on the other side, the lower K implies a flamelet that extends over a bigger physical space, and consequently is less sensitive to the filtering operation. Altogether this reduces the chemical source term to almost one fourth of its value, the progress variable correction terms drop down to one sixth, and an bigger cut down is exerted over the mixture fraction correction terms which decrease between 20 and 30 times from $x = 1$cm to $x = 4$cm. All of which is coherent with the flame profile modification only at the first upstream section.

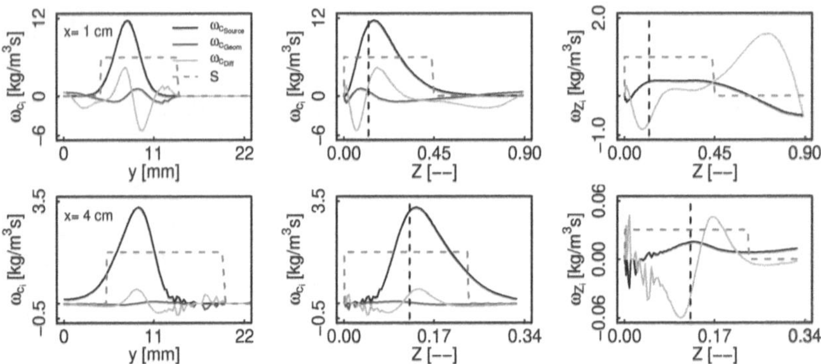

Figure 4.15: Progress variable and mixture fraction source terms along two different radial sections for $\Delta = 4$mm. Solid lines source terms, red dashed line flame sensor, black vertical dashed line Z_{st} location.

Focusing on the sensor, its outermost limit on the radial direction reaches the zone where all the correction terms extinguish, while this property does not hold for the innermost limit. When observing this behavior in mixture fraction space at the center column, it can be appreciated that indeed the sensor goes from $Z = 0$ to a varying upper limit. For the mixture fraction correction terms shown in the right column, it is so that at $x = 1$cm the highest values are located towards the fuel side, and indeed lie in the deactivated sensor zone. This behavior changes at the second radial position, where despite the already mentioned source terms decrease, the highest values lie within the sensor limits.

In order to justify these neglected Z terms at the first coordinate, it is of paramount importance to comprehend S behavior. It has already been pointed out based on Figure 4.12 that the one-dimensional approximation holds quite well close to the reaction zone, and the same conclusion is reinforced by observing S activation on Figure 4.15, i.e. the simulation gradients match those of the manifold flamelets. Close to the centerline the phenomenon differs considerably from the counterflow condition, then due to the downstream fuel consumption the gradients in the axial direction play a non-negligible role, and so the one-dimensional flamelet hypothesis loses its validity. Considering the already described K decreasing curve from Z_{max} towards c_{max}, high K values imply high Z gradients as well as big correction terms. In this context, S determines the zone where the counterflow flamelet indeed conforms with the simulated condition, and out of which the correction terms have no physical meaning.

When considering the whole sensor range plotted in mixture fraction space, in addition to the decrease due to Z decay along the centerline, the above discussed radial offset owing the non-holding of the counterflow hypothesis on the fuel side can be clearly seen on figure 4.16. The distance on the physical space is nearly the same for the first two sections and slightly decreases for the third one. In this last one the fluctuating S has not been taken into account, as it obeys oscillations in K

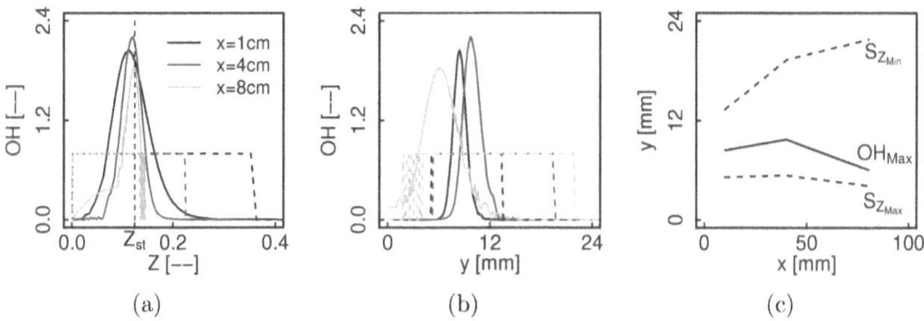

(a)   (b)   (c)

Figure 4.16: OH radial profiles for $\Delta = 2$mm at different downstream locations represented in mixture fraction (4.16a) and physical space (4.16b). Figure 4.16c corresponds to the limit values of Figure 4.16b along the axial direction. Solid lines OH, dashed lines flame sensor.

close to the equilibrium as seen in Figure 4.12 and not a physical reason. Figure 4.16b shows the expected increase in OH profile width as K decreases, while the adequate sensor response to this thickening through the widening of S limits it to be appreciated on Figure 4.16c.

Aiming to further illustrate S behavior as the flow evolves downstream, Z gradients in the radial and axial directions are then presented in Figure 4.17. It can be seen that qualitatively the trend between the three radial sections does not change. There is a significant axial contribution at Z_{max} which decreases in the oxidizer direction. The sensor describes a constant profile except for the last coordinate, where the already explained intermittent behavior is observed. The employed sensor definition proves not only to detect the counterflow regions, but it shows that by activating the FTACLES model in the regions where the criteria holds the underlying physics behind flame filtering can be correctly represented.

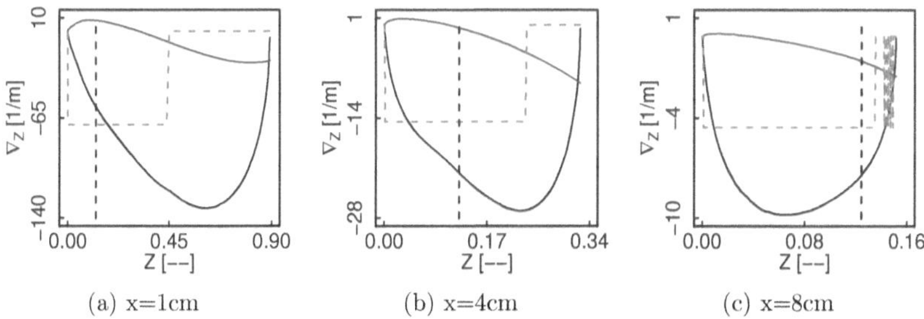

(a) x=1cm   (b) x=4cm   (c) x=8cm

Figure 4.17: Mixture fraction gradient radial profiles in mixture fraction space for $\Delta = 4$mm at different downstream locations. Black solid line radial gradient, blue solid line axial gradient, and the dashed line corresponds to the flame sensor.

4.4　Perspectives: differential diffusion effects

This section sets a starting point for the consideration of differential diffusion effects in the non-premixed filtered tabulated chemistry formalism. First the study case is solved with a detailed chemistry solver employing two distinct diffusion models. The results are compared in order to illustrate the effect of preferential diffusion on the current configuration. Subsequently different strategies to include these effects on flamelet-based approaches are presented. The last subsection briefly discusses how specific features of these preferential diffusion models could be coupled with the non-premixed filtered tabulated chemistry model and guidelines are proposed.

4.4.1　Preferential diffusion effect on the study case

In order to appreciate the role of preferential diffusion on the considered configuration, the case was solved employing the detailed chemistry solver *laminarSimpleSMOKE* with two different diffusivity models. One simulation was performed employing $Le = 1$ assumption, now DC1, and a second simulation was done with the multi-component model, now DC2. Figure 4.18 presents the mixture fraction and temperature radial profiles at three different sections. The mixture fraction has been computed from the element mass fractions using Equation 2.77. Both Z and T profiles converge at $x = 1cm$, a slightly faster Z decay for DC2 can be observed from $x = 4cm$ and at $x = 8cm$ DC2 maximum temperature presents a reduction with respect to DC1.

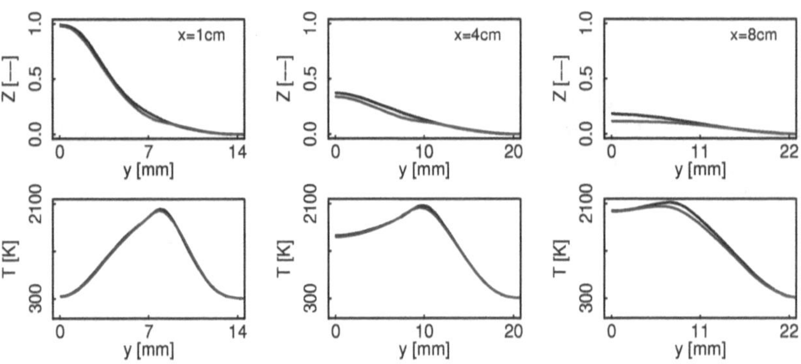

Figure 4.18: Radial profiles of mixture fraction and temperature for detailed chemistry simulations with different diffusion modeling. Black color $Le = 1$, blue color multi-component diffusivity model.

Figure 4.19 depicts different species radial evolution. As expected H_2 evolution changes considerably when the constraint of the unity Le is relaxed. It diffuses much faster, therefore the peak value decreases and the profile becomes much more homogeneous. For H_2O and CO, with average Le of 0.854 and 1.171 respectively, the deviations are less pronounced at the first two sections, though they become clearly relevant at $x = 8cm$. The OH peak concentration is slightly under predicted though the profile shape appears to be less affected than the other considered species.

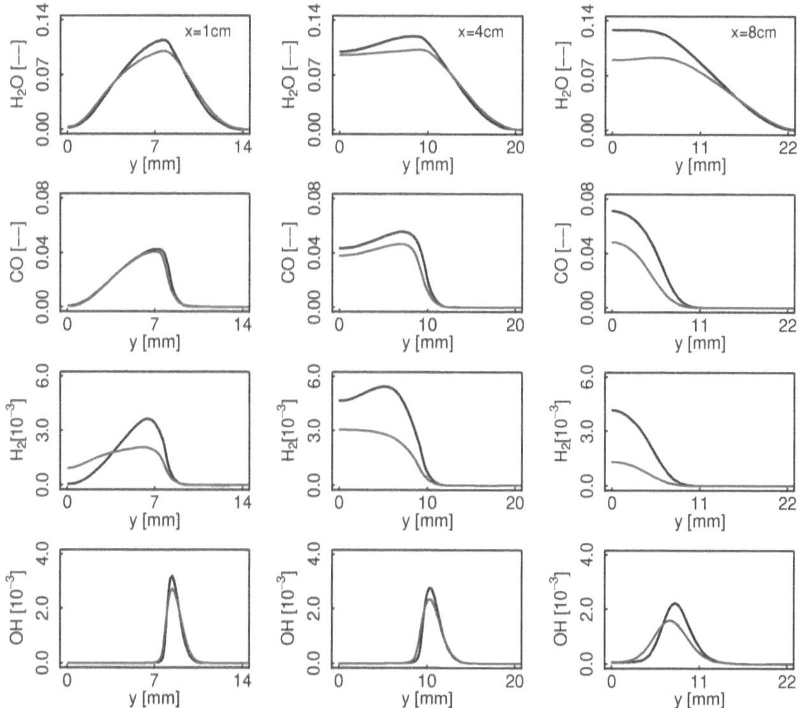

Figure 4.19: Species radial profiles for detailed chemistry simulations with different diffusion modeling. Black color $Le = 1$, blue color multi-component diffusivity model.

4.4.2 Differential diffusion modeling for flamelet methods

Given the relevance of the problem, distinct approaches to deal with differential diffusion for flamelet-based methods have been proposed in literature. Two features that can be employed for classification purposes are: whether the flamelet equations are solved in physical space as in the FGM methodology[142] or in mixture fraction space using a coordinate transformation as in the flamelet concept[110], and second, the way the passive scalar Z is defined. Acknowledging that the non-premixed filtered tabulated chemistry approach performs the filtering operation in physical space, special emphasis will be given to modeling strategies developed in this context. Nonetheless, at the end of the section one model developed for flamelet equations in Z space will be briefly described due to the pertinence of employed concepts.

The flamelet equations in FGM context were already introduced in section 3.1. For the sake of clarity the species and enthalpy equations are rewritten here

$$\frac{\partial \rho u Y_i}{\partial s} + \rho K Y_i = \frac{1}{Le_i} \frac{\partial}{\partial s} \left(\frac{\lambda}{c_p} \frac{\partial Y_i}{\partial s} \right) + \dot{\omega}_{Y_i} , \tag{4.2}$$

$$\frac{\partial \rho u h}{\partial s} + \rho K h = \frac{\partial}{\partial s}\left[\frac{\lambda}{c_p}\frac{\partial h}{\partial s} + \sum_{1=1}^{N_s} h i \frac{\lambda}{c_p}\left(\frac{1}{Le_i} - 1\right)\frac{\partial Y_i}{\partial s}\right]. \tag{4.3}$$

Recalling that the progress variable results from a linear combination of species

$$c = \sum_{i=1}^{N_c} \alpha_i Y_i, \tag{4.4}$$

where N_c corresponds to the number of species considered for c, it can be seen that $Le_i \neq 1$ impacts both h and c conservation.

In the FGM methodology the equations are solved in the physical space, thus contrary to the laminar flamelet approach by Peters[110], the mixture fraction is not necessary for the flamelet system solution but rather for the manifold parameterization. It follows that different definitions for Z have been employed in the literature.

FGM-NUnLe1

In the analysis of the FGM capability to describe non-premixed flames[144], here referred to as FGM for non-unity Lewis number 1 (FGM-NUnLe1), Z is computed as the transport of a passive scalar without preferential diffusion effects

$$\frac{\partial \rho u Z_t}{\partial s} + \rho K Z_t = \frac{\partial}{\partial s}\left(\frac{\lambda}{c_p}\frac{\partial Z_t}{\partial s}\right). \tag{4.5}$$

Considering a two-dimensional parameterization (Z, c), differential diffusion effects attain uniquely the progress variable. The transport equation reads

$$\frac{\partial (\rho c)}{\partial t} + \frac{\partial}{\partial x_i}(\rho u_i c) = \frac{\partial}{\partial x_i}\left(\frac{1}{Le_c}\frac{\lambda}{c_p}\frac{\partial c}{\partial x_i}\right) + \dot{\omega}_c, \tag{4.6}$$

and

$$\frac{1}{Le_c} = \sum_{i=1}^{N_c}\left(\alpha_i \frac{1}{Le_i}\frac{\partial Y_i}{\partial s}\right)\left(\frac{\partial c}{\partial s}\right)^{-1}, \tag{4.7}$$

where the parameter $\frac{1}{Le_c}$ can be computed in the preprocessing step and is thus stored in the look-up table.

The method appears to be partially capable of reproducing differential diffusion effects. The formulation satisfactorily describes planar chemical structures, while the results on a 2-D coflow laminar diffusion flame appear to be more modest. The authors attributed the discrepancies, mainly arising near the axis, to the flame curvature, as these effects are not accounted for in the laminar flamelets employed for the manifold generation[144].

FGM-NUnLe2

The study considering the inclusion of differential diffusion effects on FGM for the modeling of stratified premixed cooled flames[41], here FGM-NUnLe2, employs a different approach as it computes Z from the combination of C,H and O element mass fractions following Equation 2.77. This definition implies that preferential diffusion modifies Z as well. It results convenient to define a generic control variable $C_k = \{C_1, C_2, ..., C_{N_{CV}}\}$, where N_{CV} the number of control variables, $C_1 = c$, and further variables e.g. Z or h are added subsequently. Then considering a parameterization $Y_i = f(C_k)$, and assuming that locally the controlling variables are function of c, i.e. $\forall\, k \neq 1 : C_k = C_k^{1D}$, the conservation equation for any control variable can be expressed employing the same structure

$$\frac{\partial\left(\rho C_k\right)}{\partial t} + \frac{\partial}{\partial x_i}\left(\rho u_i C_k\right) = \frac{\partial}{\partial x_i}\left(\frac{\lambda}{c_p}\frac{\partial C_k}{\partial x_i}\right) + \frac{\partial}{\partial x_i}\left(d_{C_k}\frac{\partial C_1}{\partial x_i}\right) + \dot{\omega}_{C_k}\,, \tag{4.8}$$

where the second RHS term describes the differential diffusion. The differential diffusion coefficient for C_k is expressed as

$$d_{C_k} = \frac{\lambda}{c_p}\sum_{i=1}^{N_s}\left[\left(\frac{1}{Le_i}-1\right)D_{k,i}\sum_{i=1}^{N_{CV}}\frac{\partial Y_i}{\partial C_k}\frac{\partial C_k^{1D}}{\partial C_1}\right]\,, \tag{4.9}$$

where $D_{k,i}$ varies for each control variable, it is precomputed and saved in the lookup table. Details on the model derivation and $D_{k,i}$ construction can be found in [41].

Equation 4.8 is a generalization of the FGM-NUnLe1 model , s.t. if only one control variable is employed, i.e. $C_k = c$, Equation 4.6 is retrieved. Moreover, this approach presents the particularity that for $k > 1$ the differential diffusion contribution on Equation 4.8 is expressed in terms of c, therefore it is explicitly added as a source term.

The approach was appraised on a 2D stratified laminar CH_4-Air flame with heat loss to the wall. The method delivers satisfactory results, however two observations can be done: though the diffusion effects are well reproduced, they are not strong in CH_4 flames. Second, the differential diffusion inclusion on transport equations during the flame simulation does not provide significant benefits, being sufficient to consider this aspect for the manifold construction, without including the extra terms of the FGM equations during run-time[41].

Consistent flamelet formulation

For flamelet equations solved in Z space, the consistent flamelet formulation for non-premixed combustion considering differential diffusion[118] defines Z from the solution of a conservation equation with an arbitrary diffusion coefficient, e.g. $Le = 1$, and appropriate boundary conditions, s.t. the passive scalar is not directly related to any combination of the reactive scalars. The model particularity relies on the fact that the species and energy conservation equations are written without any assumption about Le, therefore employing more complex diffusion models than Fick's law. The study demonstrates the relevance of diffusion modeling, for instance through the computation of binary coefficients and the correction velocity. The model has been conceived for a (Z, χ) coupling with the flow field, hence if a tabulation strategy considering c is used, e.g. FGM or FPV approach, its transport considering the relaxation of Le assumption becomes a not straightforward problem.

4.4.3   Further research possibilities

Turbulent combustion simulation with LES coupled to an FGM approach neglecting preferential diffusion effects, e.g. the turbulent lifted H_2/N_2 jet flame in vitiated co-flow study in [10] or the the stability effects of H_2 diluted into CH_4 in a real industrial combustor analysis in [101], continues to be a common practice despite the awareness of preferential diffusion effects under such conditions. The FGM approaches above described, to the best of the author's knowledge have not found widespread utilization, despite the promising preliminary outcome. Given the encouraging results obtained with the non-premixed filtered tabulated chemistry approach, the inclusion of preferential diffusion within the formalism appears to be a worthwhile alternative to fill this gap. It follows that two main research orientations are then proposed: the first one concerning the analysis of laminar non-premixed coflow flames, and the other tackling turbulent diffusion conditions.

Model extension to more complex flamelet configurations

It has been demonstrated that planar counterflow flames are not able to correctly reproduce the results in laminar non-premixed coflow flame configurations, especially along the centerline since curvature effects play a fundamental role. In order to retrieve a detailed chemistry solution considering differential diffusion effects with an FGM approach, the manifold should be constructed employing more complex non-premixed flamelet configurations. For instance, recently the multistage stage FGM method, with a manifold consisting of 1D curved flames, was used to take into account curvature-preferential diffusion interactions in the simulation of 1D laminar flames[59]. The results reveal that when the curvature effects are included in the manifold, the MuSt-FGM results agree well with the detailed chemistry results; which is not the case when the curvature effects are ignored. Since the non-premixed filtered tabulated chemistry equations have been derived based on the counterflow flamelet configuration, the method correction terms should be recomputed for distinct flamelet governing equations. In other words, the filtering principle and the closure strategy would hold, but the expression for the explicitly computed unresolved terms would vary.

The idea consists to employ an appropriate flamelet configuration, taking into account tangential effects, to describe the laminar non-premixed coflow case. Thus the non-counterflow effects will be undoubtedly separated from the filtering ones. A proposed methodology would be as follows:

- Solve the case employing the FGM manifold generated using a distinct flamelet configuration than the planar counterflow one.

- Verify that the results adequately match with the differential diffusion detailed chemistry ones.

- Filter the new FGM results.

- Solve the case with the non-premixed filtered tabulated chemistry approach, now derived for the new flamelet configuration.

- Compare the model results against the filtered FGM.

Two main observations should be made: due to the low strain rate, and thus the big flame thickness downstream a laminar coflow flame, a rather limited profile transformation resulting from the filtering operation will be expected. The second aspect concerns the main purpose of directly filtering a laminar coflow flame within this study, which does not rely on the results themselves, but on its role as an intermediate step to substantiate a given formulation for its further use on LES. If the filtered laminar flame is considered as an independent entity, its relevance is mostly theoretical, being rather difficult to justify it from a merely practical perspective. Therefore, the filtered laminar coflow flame research should be carried out hand by hand with further studies aiming to assess flame curvature contribution on real non-premixed turbulent flames, in order to determine if the additional complexity required to describe the coflow flame might payoff.

Variable Le_i consideration in counterflow flamelets

Section 4.4.2 presented two formulations for the inclusion of variable Le_i in counterflow flamelets within the FGM formulation. For a (Z, c) parameterization, as the one employed in this study, the first formulation modifies uniquely the c transport equation, while the second adjoins a differential diffusion term to Z equation as well. The second formulation has been derived considering a premixed condition, therefore it has been assumed that locally the control variables depend on c, and for this reason the diffusion term added to Z equation is expressed in terms of $\frac{\partial c}{\partial x_i}$. Though the followed approach, namely, to consider differential diffusion effects on Z behavior does seem appealing, the formulation should be carefully revised as to make it fully consistent for the non-premixed regime. It follows, that though simplified, the first formulation does not need any further modifications, and its implementation results into a relatively straightforward problem, contrary to the model expansion to other flamelet configurations discussed in the section above. The term representing the preferential diffusion, described by Equation 4.7 is tabulated in terms of the existing controlling parameters, hence it does not lead to any dimension increase.

Considering the LES framework, to assure the counterflow flamelets adequate performance, the case should be chosen so that the steady laminar flamelet assumption holds, as it is defined for the diffusive turbulent combustion regime diagram. A possible approach would be to initially assess the model performance with the first formulation, in order to characterize the sensitivity to preferential diffusion. Based on the results, more sophisticated strategies could be developed, for instance considering a modified Z equation, or even expanding the flamelet diffusion modeling to more complex expressions than Fick's law.

4.5 Conclusion

This chapter assessed the performance of the non-premixed filtered tabulated chemistry formalism on a laminar coflow diffusion flame. The model appeared to be considerably sensitive to the flame dimensionality. Unphysical profile distortions arised when points described by highly strained regions of the manifold, and consequently having big model correction terms, did not comply with the one-dimensional flamelet hypothesis. Aiming to verify the soundness of this assumption a flame sensor based on the mixture fraction gradient, with a tolerance to take into account the numerical grid resolution, was introduced.

Simulations performed using the model coupled with the flame sensor were compared against the direct filtering of a reference solution, and good agreement was obtained for the laminar flame. This is a fundamental result as it demonstrates the capability of the formalism to preserve the flame characteristic features. In this way the sensor determined model activation adequately represents the underlying physics behind flame filtering and so it endorses the consistency of the numerical procedure as a thickening operation.

Expanding on the findings, using three different filter sizes, it turned out that:

- Far from the centerline the flame front does not only satisfy the counterflow hypothesis, but K remains unaltered after the filtering operation, what can be understood as the filtering of the same trajectory, namely of the same flamelet.

- The filter effect at the radial reaction zone is then transported throughout the domain and is the cause for the centerline profile modifications, where the model is not active.

- The profile extension due to the filtering process is mainly driven by the model correction terms, while the decrease in the peak values of non monotonically evolving variables depends on the filtered parameters that enter the transport equation.

- The sensor adequately identifies the multidimensional effect at the centerline, and its active range both in Z as in physical space changes throughout the domain, in accordance with the strain rate decrease and higher flame thickness.

The present study sets a milestone towards the application of non-premixed filtered tabulated chemistry as a turbulent combustion modeling strategy in the LES framework. The promising results obtained on the coflow laminar diffusion flame encourage the further model appraisal on a more complex turbulent configuration.

Chapter 5

Turbulent cases: flames D and E

The non-premixed FTACLES is based on the idea that in turbulent diffusive combustion regime, if there is no SGS wrinkling, the thin flame front can be amplified as to be resolved on practical LES numerical grids. In order to do this, one-dimensional counterflow flamelets are directly filtered, e.g. with a Gaussian operator, and the resulting manifold is parameterized in terms of filtered control variables, i.e. $\widetilde{Z}$ and $\tilde{c}$. The modified flame structure is now coherent with the domain discretization, and therefore the retrieved variables, e.g. the source term, can be directly employed on each computational cell.

Such weakly wrinkled conditions can be found for example in liquid rocket engines operating at elevated pressures. The simulation of such systems is highly complex, both concerning the underlying physics description as from the computational resources point of view. For a vast majority of practical turbulent combustion systems however, the flame structure is distorted by the turbulence structures and SGS effects cannot be neglected. Numerous experimental studies have been carried out for these conditions, offering abundant high quality data for model validation purposes.

The question that this chapter aims to answer reads: Can the coupling of the non-premixed FTACLES with a SGS modeling function adequately describe a wrinkled flame front condition? The problem is of high scientific relevance, as satisfactory results would suggest the possibility of SGS closure in diffusive combustion based on filtering arguments rather than on a statistical basis. In order to tackle this bivariate problem, the attention is focused on the non-premixed FTACLES model performance. Due to the novelty of the approach, i.e. thickening strategies are well known for premixed combustion but to the authors knowledge no models exist for diffusive combustion, no SGS functions have been developed to describe the modified turbulence-chemistry interaction of a thickened diffusion flame. Therefore, the SGS contribution is computed employing an expression based on the efficiency function ξ proposed in [29],whose good performance has already been demonstrated.

As application object, the Sandia flames D and E have been selected, as they are well documented in terms of experimental data and model validation in the literature. A non-exhaustive review of employed combustion models in LES context includes FPV[67][40], CSE [51], ESF [73][155], and Lagrangian approaches [119] [55]. Flame D has been studied employing premixed and non-premixed generated manifolds[148] and has also served to address further phenomena such as pollutant formation[72] or turbulence-radiation interactions[100].

The chapter's structure is then as follows:

- The flames D and E are introduced. Flame D is simulated employing the well known FPV model in Section 5.1. The case settings and the mesh resolution are thus verified. Satisfactory results to be used as reference for the subsequent analysis are gathered.

- Section 5.2 is devoted to the comparison of the premixed and non-premixed FTACLES without consideration of the SGS wrinkling. The goal is to assess the analogousness of the approaches, as to further justify the incorporation of the efficiency function for the modified turbulence-chemistry interaction in the non-premixed case.

- The main objective of the chapter, i.e. the appraisal of the non-premixed FTACLES model coupled with the proposed ξ is presented in Section 5.3.

- In order to gain further insight on the model performance, a sensitivity analysis with respect to the filter size is carried out in Section 5.4. The filter effects on the species prediction as well as the change in turbulence interaction are thoroughly adressed.

- Section 5.6 presents the main conclusions and findings.

5.1   Case description

For the reasons evoked above, the piloted partially premixed methane/air Sandia flames D and E configuration[8] is selected and simulated. The main jet consists of a mixture 25%CH_4 and 75%Air by volume, entering through a pipe with $D = 7.2$mm. The pilot stream has a diameter $D = 18.2$mm, an equivalence ratio of $\phi = 0.77$ and a temperature $T = 1880K$. The unconfined system ends with an outer air coflow at $T = 291K$. The fuel jet bulk velocity is $U_{Fuel} = 49.6$m/s and $U_{Fuel} = 74.4$m/s, and the pilot velocity $U_{Pilot} = 11.4$m/s and $U_{Pilot} = 17.1$m/s, for flames D and E respectively. Therefore the former presents a very small probability of local extinction which increases for the latter. Due to the high mixing rates combustion takes place in diffusive regime, with the reaction zone around the stoichiometric value. The statistical errors of the measurements for the mean major species and temperature have been reported to be below 5% and those for CO are below 10%[66].

The CFD code OpenFOAM is used with a low-Mach pressure-based approach. The combustion is addressed employing various tabulated chemistry models. The system can be considered adiabatic, therefore the thermophysical state can be fully described by means of Z and c, and no energy equation is transported. Different solvers, i.e. *lowMachNonPremFTACLESfoam*, *lowMachPremFTACLESfoam* and *lowMachFPVfoam*, depending on each specific combustion model are employed, hence the exact formulation of Z and c equations varies as a function of the closure approach. Altogether six equations are solved, three for the momentum, and one for p, Z and c respectively. In addition the mixture fraction variance Z_v^2 is also solved on the presumed β-PDF model as below specified. Turbulence generation at the inlet is based on the digital filter method by Klein et al. [75]. The WALE model [104] is used for the subgrid scale turbulence while the SGS scalar fluxes are closed either with the eddy diffusivity model or with the explicit FTACLES model terms. The wrinkling effect is taken into account as explained in section 3.1. ct was set equal to 0.85 for the flame sensor. The progress variable is defined as $c = \sum Y_i / \sum Y_{i,eq}$, and the selected species i are CO_2 and CO as proposed in [52]. The employed boundary conditions are presented in Table 5.1.

Table 5.1: Flames D and E boundary conditions. z.G stands for zero-gradient.

Stream	Z	c	U (m/s)	p (Pa)
Fuel	1	0	U_{Fuel}	z.G
Pilot	0.276	0.7042	U_{Pilot}	z.G
Coflow	0	0	0.9	z.G
Wall	z.G	z.G	0	z.G
Side	z.G	z.G	pressureInletOutletVelocity	totalPressure
Outlet	z.G	z.G	pressureInletOutletVelocity	totalPressure

The three-dimensional numerical grid is conical, non uniform, structured with an O-grid, and it consists of a pipe $L_P = 13D$ and a combustion chamber $L_{ch} = 80D$. Two different resolutions were used in this study, as described in Table 5.2. The mesh M_2 results from a systematic coarsening of mesh M_1, with a factor close to 1.5 in all directions, which leads to a reduction of about 2.5 the number of cells.

Table 5.2: Numerical grid parameters. n is the number of divisions, the subindex ch stands for chamber, p for pipe, r for radial, and L for axial direction.

Code	$n_{r,ch}$	$n_{r,p}$	n_ω	$n_{L,ch}$	$n_{L,p}$	Cells
M_1	143	24	17	394	145	4Mi
M_2	108	15	11	329	100	1.6Mi

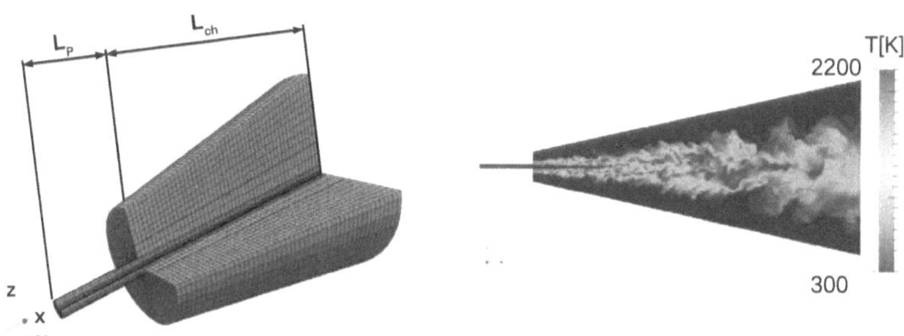

Figure 5.1: Numerical mesh and temperature field for Flame D.

The simulations were done on the Mother High Performance Computer of the Institute of Energy and Power Plant Technology at the Technical University of Darmstadt for the phase 3 infrastructure. A decomposition into 96 processors, was used for M_1 and 48 processors for M_2. After a statistically stationary flow was obtained, the results were averaged for 0.11s, for which the study in [66], using the same domain length of $80D$, proved the statistics convergence.

The model performance was appraised employing M_1. Simulations were done employing the non-premixed FTACLES, FPV and purely FGM, i.e. no model. The sensitivity analysis on Flame D with respect to the filter size effect was carried out employing M_2 for computational time convenience. Table 5.3 describes the considered scenarios, and for each one of them three simulations were performed varying the filter size $\Delta = 2, 3, 4$ mm respectively. The numerical grid satisfies that the cell size is smaller than the filter size all along the reactive part of the flame, that is, where the model is active. In total 14 simulations were performed with M_2: 6 employing the non-premixed FTACLES, 6 with the premixed counterpart, one with the FPV model and one with purely FGM.

Table 5.3: Cases simulated with M_2.

Code	Combustion model	Flamelets	ξ	Number of simulations
FTNP1	FTACLES	Non-Premixed	yes	3
FTNP2	FTACLES	Non-Premixed	no	3
FTP1	FTACLES	Premixed	yes	3
FTP2	FTACLES	Premixed	no	3
FPVNP	FPV	Non-Premixed		1
FGMNP	Pure flamelets	Non-Premixed		1

The non-premixed manifold is generated using the FGM method with counter-flow flamelets. The strain rate has been varied from $1s^{-1}$ up to $1082s^{-1}$ which corresponds to the last ignited flamelet, and then $1083s^{-1}$ corresponds to a pure mixing flamelet. Time-dependent flamelet solutions have been employed to fully span the composition space between the last ignited flamelet and the pure mixing line. The flamelets have been generated employing the detailed chemistry solver CHEM1D[136], where the $GRI - 3.0$[134] detailed chemical mechanism has been considered with $Le = 1$ assumption.

In order to verify the mesh adequacy, as well as the inlet parameters for the

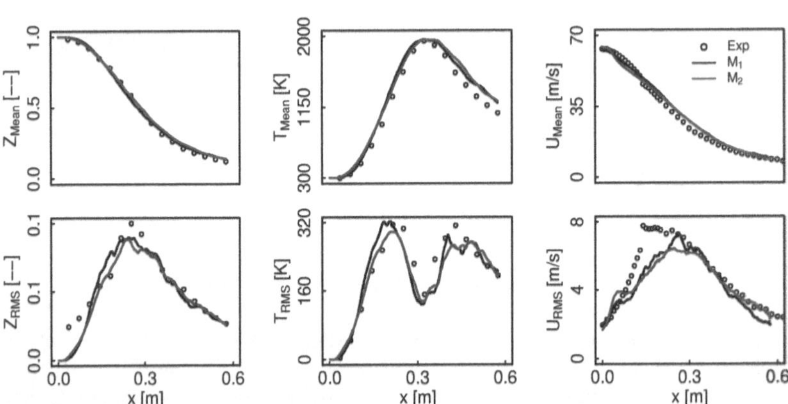

Figure 5.2: Flame D mixture fraction, temperature and velocity profiles along the centerline for the FPV model with two different numerical grids.

turbulence generator, Flame D is initially investigated using the FPV model, before assessing the non-premixed FTACLES model. In this approach the turbulence chemistry interaction is taken into account by means of a joint subgrid PDF of Z and c. A presumed $\beta = f(Z, Z_v^2)$ is chosen for the former, while the conditional subgrid PDF $P(c|Z)$ is modeled employing a δ function as in [114]. Hence additional to the six above mentioned transported variables the variance of the mixture fraction Z_v^2 is solved as well. Figure 5.2 presents the obtained results which match quite well the experimental data, in terms of mean and RMS values.

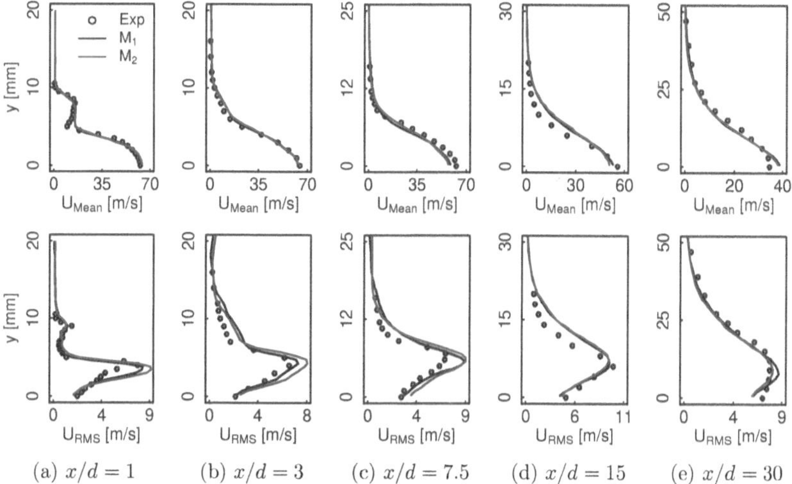

Figure 5.3: Flame D FPV model velocity mean and RMS radial profiles for two different numerical grids.

For the turbulence generator a power law profile has been given on the fuel, and subsequently the turbulence intensity as well as some model specific parameters like the grid plane divisions and the lengthscales ratio have been calibrated. Figures 5.3 and 5.4 present the velocity and mixture fraction results. These results correctly match the experiments, and the jet spreading rate is adequately predicted.

The temperature profiles are compared in Figure 5.5 where the experimental data is properly predicted. The numerical grid and the case settings have been successfully validated. The FTACLES approach can now be assessed.

5.2   Premixed vs non-premixed

Aiming to further assess their equivalence, the achievements with the two model approaches have been appraised on Flame D in the absence of SGS wrinkling. Employing the nomenclature introduced in table 5.3 this corresponds to cases FTP2 and FTNP2. This expands the manifold transformation analysis presented in Section 3.3 as it compares the two formalisms as a whole. A similar system response would further suggest the use of a wrinkling function in the non-premixed case in analogy to the premixed approach.

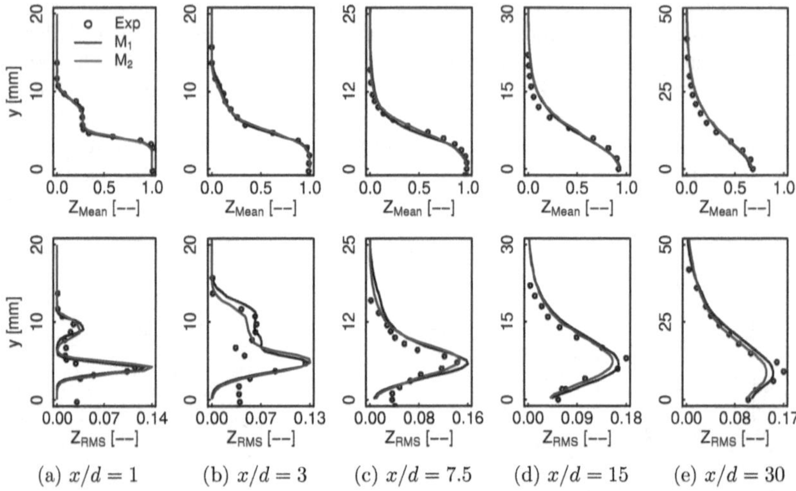

(a) $x/d = 1$ (b) $x/d = 3$ (c) $x/d = 7.5$ (d) $x/d = 15$ (e) $x/d = 30$

Figure 5.4: Flame D FPV model mixture fraction mean and RMS radial profiles for two different numerical grids.

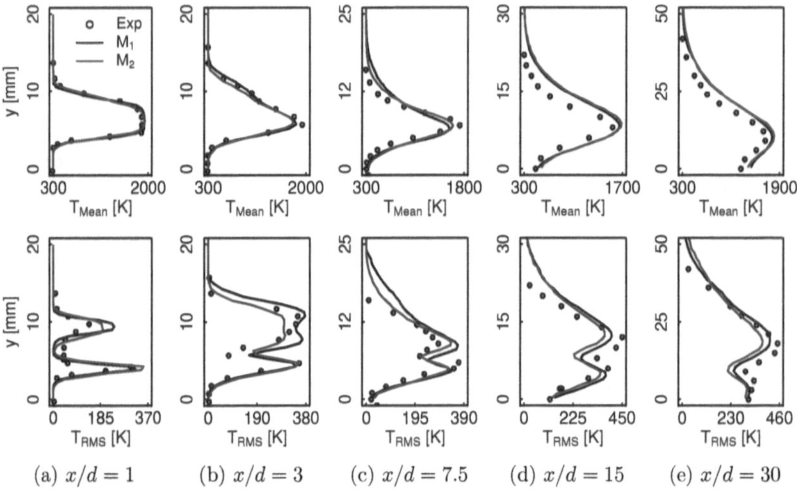

(a) $x/d = 1$ (b) $x/d = 3$ (c) $x/d = 7.5$ (d) $x/d = 15$ (e) $x/d = 30$

Figure 5.5: Flame D FPV model temperature mean and RMS radial profiles for two different numerical grids.

5.2.1 General features

Not considering the modified turbulence structures interaction distorts Z and U evolution close to the axis. A faster profile decay can be identified along the centerline, while just minor differences are observed along the radial sections. Even though the models differ in the way Z and c transport are solved, given that Z approximately coincides in both cases, the analysis will focus on the reaction prediction. The latter appears to be strongly impacted by the filter size as depicted in Figure 5.6.

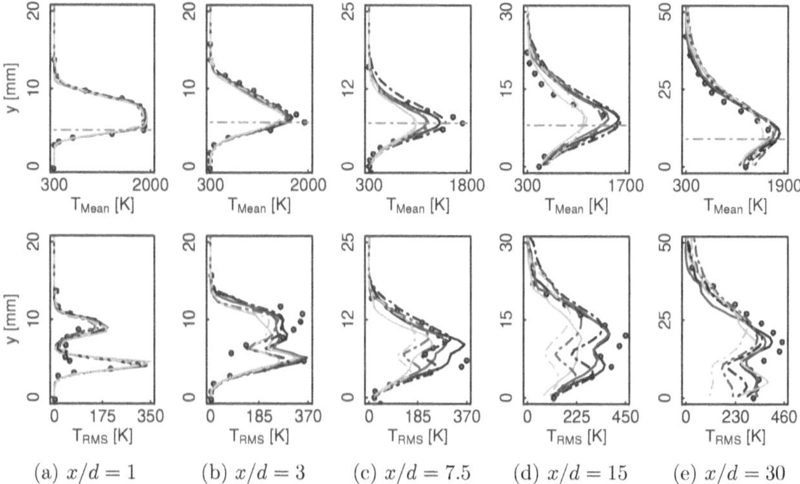

(a) $x/d = 1$ (b) $x/d = 3$ (c) $x/d = 7.5$ (d) $x/d = 15$ (e) $x/d = 30$

Figure 5.6: No ξ premixed and non-premixed FTACLES temperature radial profiles for three filter sizes. Solid line premixed, dashed line non-premixed. Black color $\Delta = 2$mm, blue color $\Delta = 3$mm, green color $\Delta = 4$mm, red line c_{max}, and circles the experimental data.

The results match the experiments for the first two sections, while deviations are only present for T_{RMS} at $x/d = 3$. Due to the lack of wrinkling the reaction rate is underestimated and a slower evolution is obtained, leading to a significant temperature under prediction which increases with the filter size from $x/d = 7.5$. The mean value profiles are similar for both methods at this position, while the fluctuations on the contrary are always less for the non-premixed, independent of the filter size. At $x/d = 15$ the peak mean values relatively coincide, though differences start to be recognizable mainly on the decay towards the oxidizer side. This difference increases further at the last section, where the mean premixed results match the experiments and the three filter sizes coincide, while the non-premixed profiles are still distinguishable, i.e. the profiles for $\Delta = 2$mm converge towards the experiments, and those for $\Delta = 3, 4$mm respectively, underestimate the values.

5.2.2   Effect of distinct sensor definitions

The distinguishable elements on c solution can be split into the computation of the diffusive flux on one side, and the estimation of the source term on the other. This comparison is far from being a straightforward task, not only because of the distinct significance of the correction terms, but also due to the flame sensor. A comprehensive analysis of S behavior and interpretation for the non-premixed case will be presented in 5.4.3. Nonetheless Figure 5.7 depicts the premixed and non-premixed sensor behavior, mainly to highlight the completely different phenomena that they identify and describe.

The premixed sensor was originally proposed for the ATF model, to deactivate the amplification in pure mixing zones, as it augmented the diffusivity without any counterweight from the chemical source term. On the contrary, the non-premixed S responds to the counterflow flame hypothesis, to avoid non-physical behavior due to Z correction source terms(see eq. 3.7). Thus the premixed profile presents an analogous shape to c, while the interpretation for the non-premixed depends more on the flame strain and the cell size.

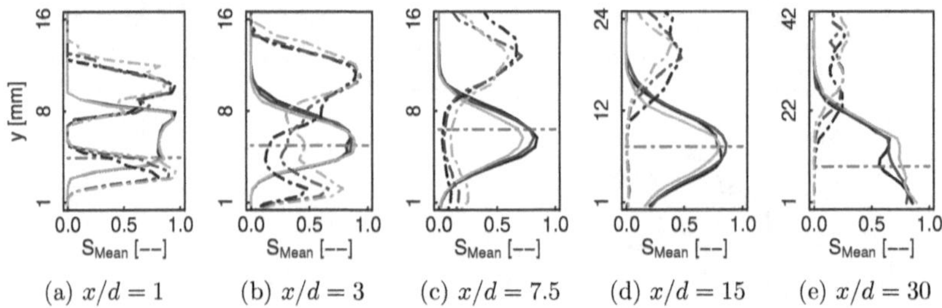

(a) $x/d = 1$ (b) $x/d = 3$ (c) $x/d = 7.5$ (d) $x/d = 15$ (e) $x/d = 30$

Figure 5.7: No ξ premixed and non-premixed FTACLES flame sensor radial profiles for three filter sizes. Solid line premixed, dashed line non-premixed. Black color $\Delta = 2$mm, blue color $\Delta = 3$mm, green color $\Delta = 4$mm, red line c_{max}.

Due to the linear correlation between c and μ, the filtered diffusivity profiles for each filter size are very similar in the two approaches and they describe the same trend observed in Figure 5.6 for T. Recalling Equation 2.71, in the premixed formalism two options appear for the model diffusivity estimation based on S. The term is computed considering the product of the filtered D with ξ and α_c for an active S, while its deactivation considers the sum of filtered and turbulent diffusivities. In the non-premixed case, Equation 3.14 multiplies the filtered D by ξ. Subsequently a deactivated S incorporates as well the turbulent diffusivity, while an active sensor will instead carry out the turbulent flux closure by means of a source term. Figure 5.8 presents the model resulting diffusivity.

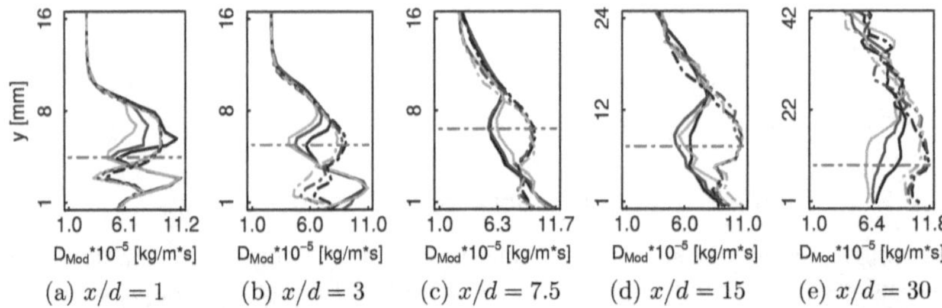

(a) $x/d = 1$ (b) $x/d = 3$ (c) $x/d = 7.5$ (d) $x/d = 15$ (e) $x/d = 30$

Figure 5.8: No ξ premixed and non-premixed FTACLES model diffusivity radial profiles for three filter sizes. Solid line premixed, dashed line non-premixed. Black color $\Delta = 2$mm, blue color $\Delta = 3$mm, green color $\Delta = 4$mm, red line c_{max}.

For the non-premixed case, the sensor has been rather introduced to render the results physically coherent (see eq. 4.1), hence it is not related to the reaction zone. Moreover, except for $x/d = 3$, the sensor tends to be deactivated in this reacting region. The model diffusivity results therefore from the sum of the filtered and turbulent components. This is opposed to the situation close to the centerline at the first two transverse sections, where S activation is responsible for the observed discrepancies with respect to the premixed results. For the premixed case on the contrary the active S term dominates in the reaction zone. The model correction coefficient α_c oscillates around unity, and generally reaches its lower value close to c_{max}, which explains the model diffusivity decrease observed at this location.

Figure 5.9 compares the filtered and final model ω_c for the premixed approach on the top row and the non-premixed one in the bottom. The non-premixed source term presents a much wider profile, as the source term expands in Z direction through the filtering operation. The premixed model substantially increases the filtered source term. By definition the premixed sensor covers the reacting zone, thus the model contributions in this region are always taken into account. The modification in the non-premixed model appears to be modest, which is directly linked to the sensor behavior.

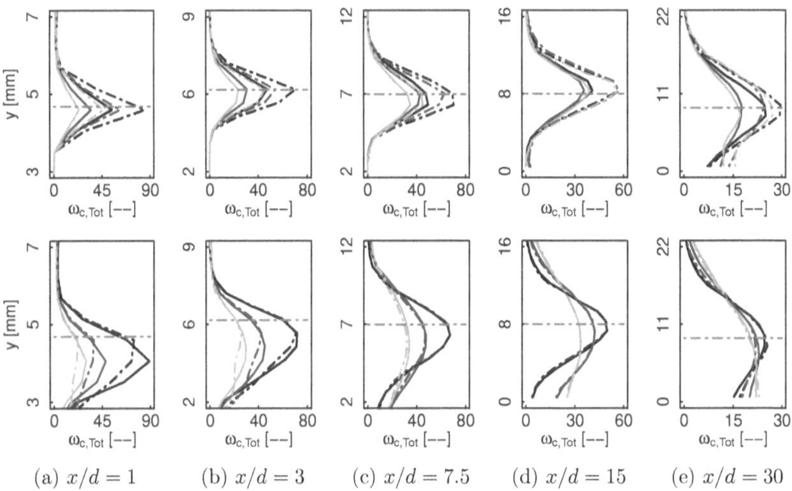

(a) $x/d = 1$ (b) $x/d = 3$ (c) $x/d = 7.5$ (d) $x/d = 15$ (e) $x/d = 30$

Figure 5.9: No ξ premixed(top) and non-premixed(bottom) FTACLES filtered and model c source term radial profiles for three filter sizes. Solid line filtered, dashed line modeled term. Black color $\Delta = 2$mm, blue color $\Delta = 3$mm, green color $\Delta = 4$mm, red line c_{max}.

The similar reaction evolution observed in Figure 5.6 takes place despite non-coincident diffusivity and source term profiles. This is a fundamental outcome, as it points to a model performance likeliness where the filter size dominates the system response. The results are quite encouraging, as they suggest that if a wrinkling function is included, the non-premixed formalism might deliver similar results as the premixed counterpart.

5.3 Assessment of non-premixed FTACLES including the SGS wrinkling

Having in mind the promising results of the previous section, the non-premixed FTA-CLES model is now assessed on flames D and E including the SGS wrinkling as expressed in equation 3.12. A three dimensional table is employed where $\varphi = f(Z, c, \Delta)$ with the filter sizes $0.5, 1, 2, 3, 4$mm, and linear interpolation is used to estimate the values among these discrete values. For Flame E the results of two numerical studies employing consolidated combustion models, namely the FPV-e (extended FPV) model[67] and the ESF (Eulerian Stochastic Fields)[13] are presented. The non-premixed FTACLES results are compared against these approaches, to highlight the capability of the formalism.

5.3.1 Validation of FTACLES vs FPV

Figure 5.10 presents the obtained mixture fraction, temperature and velocity centerline profiles for Flame D, where the FPV results are shown as well. The results adequately match the experiments, and slight differences appear between the two approaches. For instance the FTACLES model better predicts the mean temperature rise, even though Z appears to be lightly shifted, and are therefore better estimated by the FPV model. Both of the models correctly retrieve the temperature fluctuation peaks. The FTACLES model slightly shifts U_{Mean} decay, which is linked to an underprediction of the fluctuations.

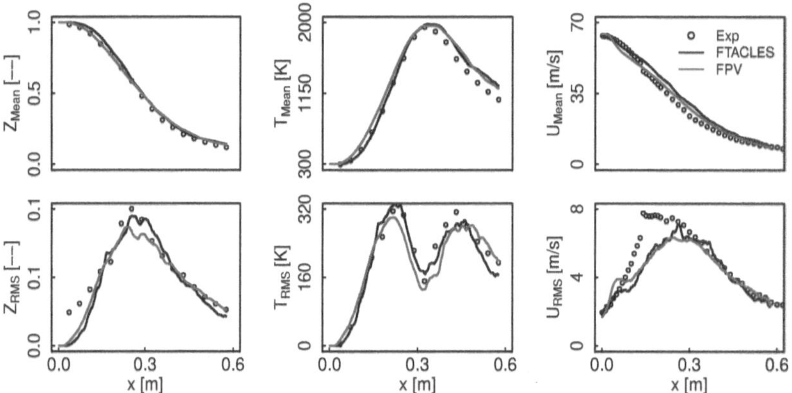

Figure 5.10: Flame D comparison between non-premixed FTACLES and FPV results for mixture fraction, temperature and velocity profiles along the centerline.

The radial profiles shown in figures 5.11 and 5.12 confirm the good model performance, for both the mean as well as the RMS values. For the non-premixed FTACLES, the coupling between the flame sensor and the model correction terms included in the mixture fraction transport equation performs adequately. The mixing process is adequately described and both models deliver analogous results.

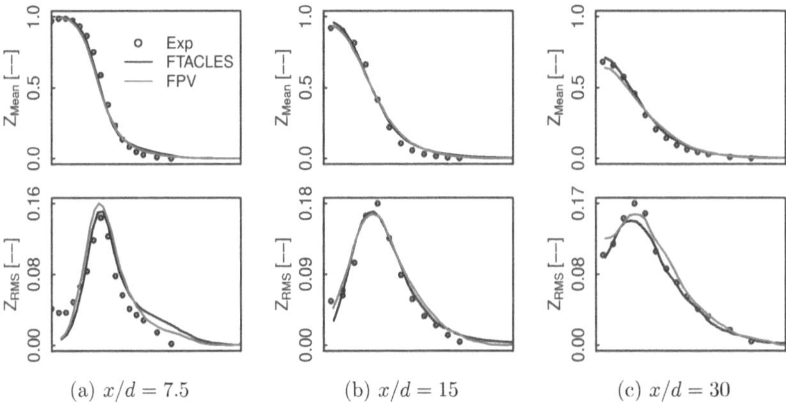

(a) $x/d = 7.5$ (b) $x/d = 15$ (c) $x/d = 30$

Figure 5.11: Flame D comparison between non-premixed FTACLES and FPV results for mixture fraction mean and RMS radial profiles.

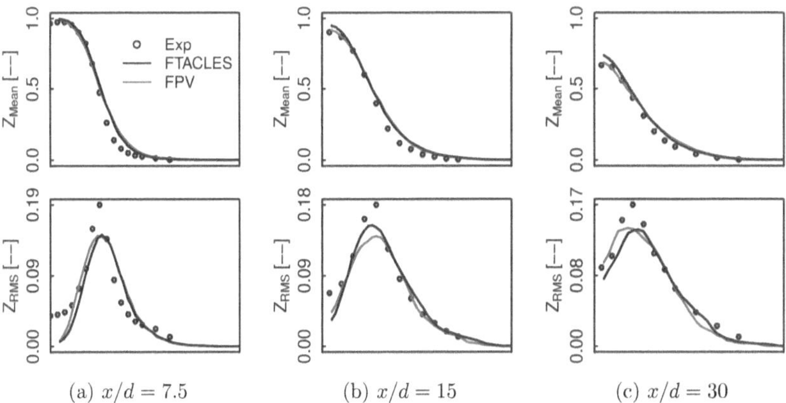

(a) $x/d = 7.5$ (b) $x/d = 15$ (c) $x/d = 30$

Figure 5.12: Flame E comparison between non-premixed FTACLES and FPV results for mixture fraction mean and RMS radial profiles.

Figure 5.13 describes the temperature behavior for Flame D. Both the mean and the RMS profiles are in good agreement with the experimental data. The non-premixed FTACLES predicts a higher fuel consumption in the first part of the domain, this results into a faster temperature rise and a better estimation than the FPV model at $x/d = 7.5$. At $x/d = 15$ the non-premixed FTACLES slightly overestimates the peak temperature and at $x/d = 30$ it outperforms the FPV model close to the centerline, while the opposite effects holds for the maximum value. During the filtering operation the higher strained flamelets undergo a bigger transformation and flamelet displacement within the manifold takes place as already explained in Section 3.2.4. The non-premixed FTACLES method adequately describes the flame physics, so that the filtered manifold is able to describe the Z, c space, hence delivering the appropriate temperature fluctuation.

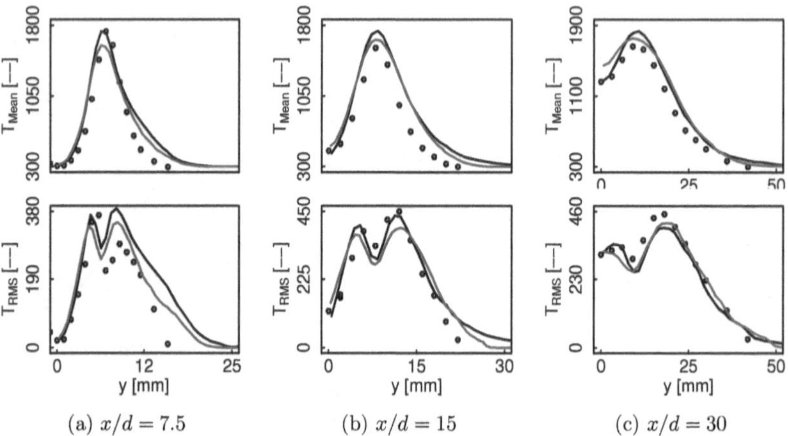

(a) $x/d = 7.5$

(b) $x/d = 15$

(c) $x/d = 30$

Figure 5.13: Comparison between non-premixed FTACLES and FPV results for temperature mean and RMS radial profiles for Flame D.

Figure 5.14 depicts the radial temperature behavior for Flame E. The FPV model(standard and extended) appears to slightly outperform the non-premixed FTACLES at the first considered transverse section. The behavior is coherent with the one described for Flame D, as the non-premixed FTACLES predicts a higher turbulence chemistry interaction in the first part of the domain. The comparison reveals an analogous mean temperature over prediction at $x/d = 15$, and afterwards all of them converge towards the experimental results at $x/d = 30$.

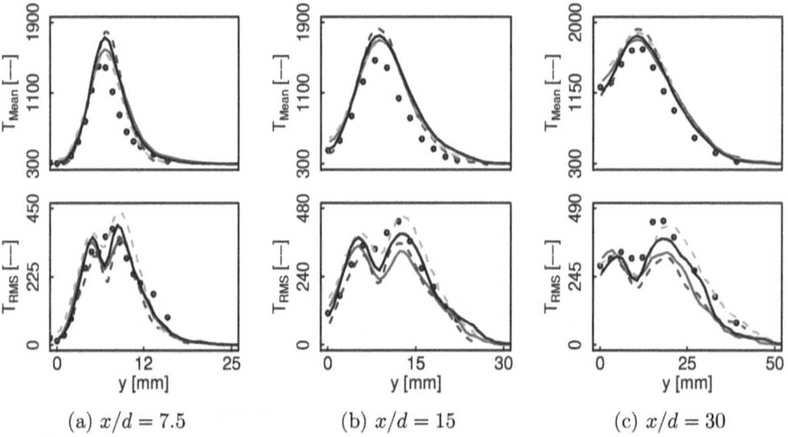

(a) $x/d = 7.5$

(b) $x/d = 15$

(c) $x/d = 30$

Figure 5.14: Comparison of temperature mean and RMS radial profiles for Flame E. Black line non-premixed FTACLES, purple line FPV, green dashed line extended FPV[67], blue dashed line ESF[13], and circles the experimental data.

5.3.2 Validation of non-premixed FTACLES: species prediction

Figures 5.15 and 5.16 present the FTACLES mean and RMS predictions for two different major species, namely CO_2 and CO. The model agrees quite well with the experimental data for Flame D. The peak values for CO mean appear to be slightly overestimated, nonetheless a very similar behavior is observed for the FPV-e results.

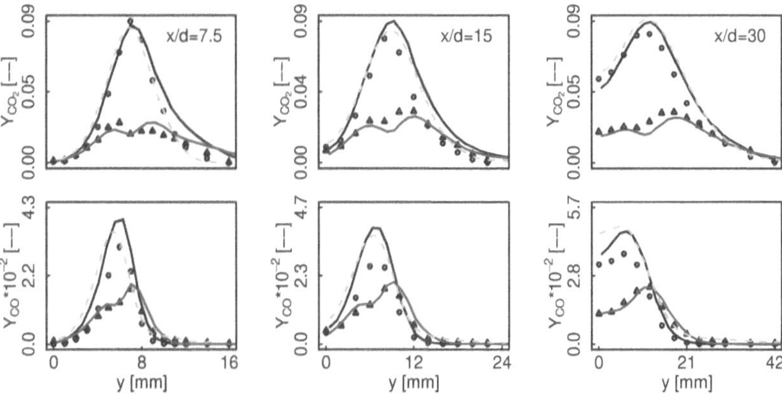

Figure 5.15: Non-premixed FTACLES CO_2 and CO mean and RMS for Flame D. Black line mean, green dashed line extended FPV[67] mean, blue line RMS, circles experimental mean, and triangles experimental RMS.

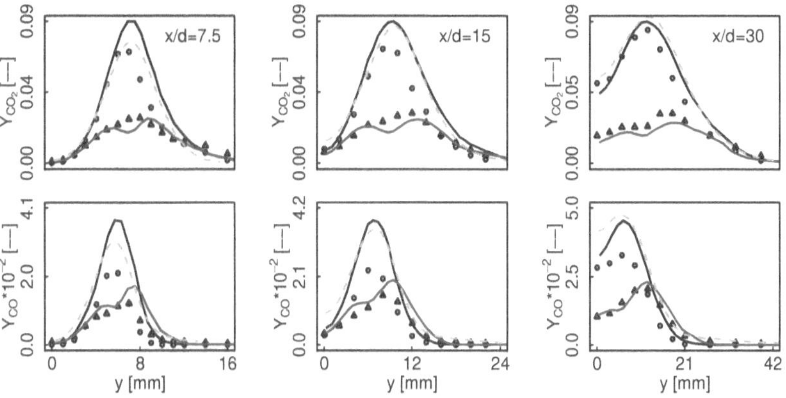

Figure 5.16: Non-premixed FTACLES CO_2 and CO mean and RMS for Flame E. Black line mean, green dashed line extended FPV[67] mean, blue line RMS, circles experimental mean, and triangles experimental RMS.

For Flame E the non-premixed FTACLES over estimates the species conversion in the initial region of the flame. The higher CO_2 and CO values at $x/d = 7.5$

are therefore coherent with the temperature behavior depicted in Figure 5.14. Both models deliver identical results at $x/d = 15$ and the non-premixed FTACLES outperforms the FPV-e at $x/d = 30$ in the radially inner region.

Figure 5.17 presents the behavior of two minor species, namely OH and H_2. Due to their considerably lower profile thickness with respect to the parameterizing variables Z, c, these variables possess a higher sensitivity to the filtering operation. By means of an adequate Z, c prediction, the FTACLES model is able to correctly estimate OH value. The model slightly over predicts H_2 mean values.

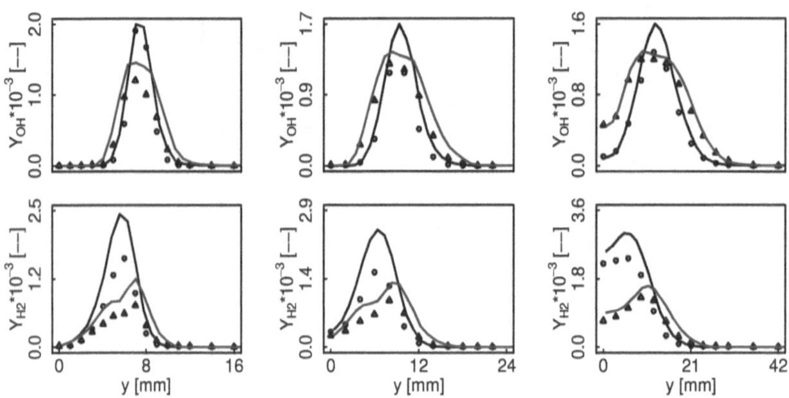

Figure 5.17: Non-premixed FTACLES OH and H_2 mean and RMS for Flame D. Black line mean, blue line RMS, circles experimental mean, and triangles experimental RMS.

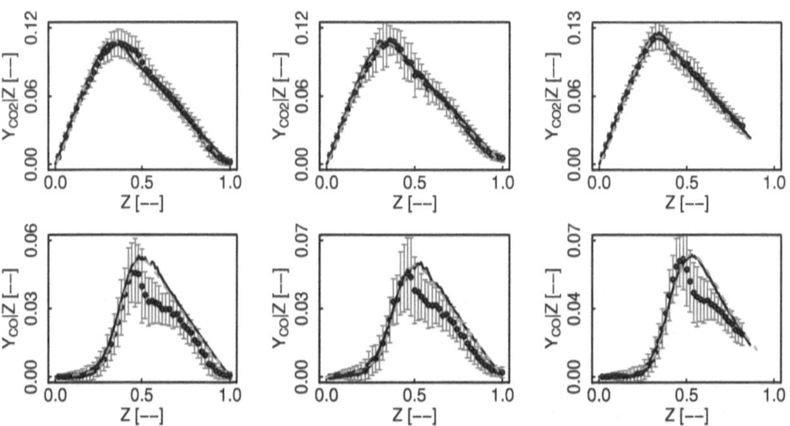

Figure 5.18: Conditional mean fractions of CO_2 and CO for Flame D. Black line non-premixed FTACLES, green dashed line extended FPV[67], circles the experimental data and error bars RMS.

In order to further appraise the non-premixed FTACLES model performance, the results of conditional averaging on the mixture fraction are presented. The model accurately predicts CO_2 on the two configurations. The CO over estimation already pointed out in Figures 5.15 and 5.16 can be identified in figures 5.18 and 5.19 as well. The comparison against the additional models depicts an similar behavior between the FTACLES and FPV-e, model, and not so distinct performance for the ESF formalism for Flame E.

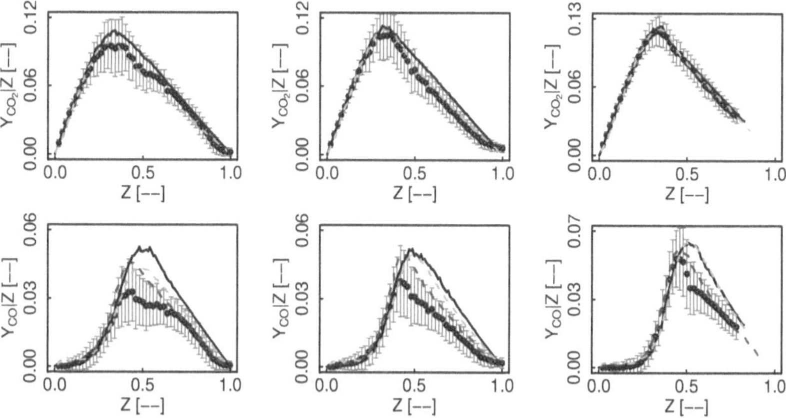

Figure 5.19: Conditional mean fractions of CO_2 and CO for Flame E. Black line non-premixed FTACLES, green dashed line extended FPV[67], blue dashed line ESF[13], circles the experimental data and error bars RMS.

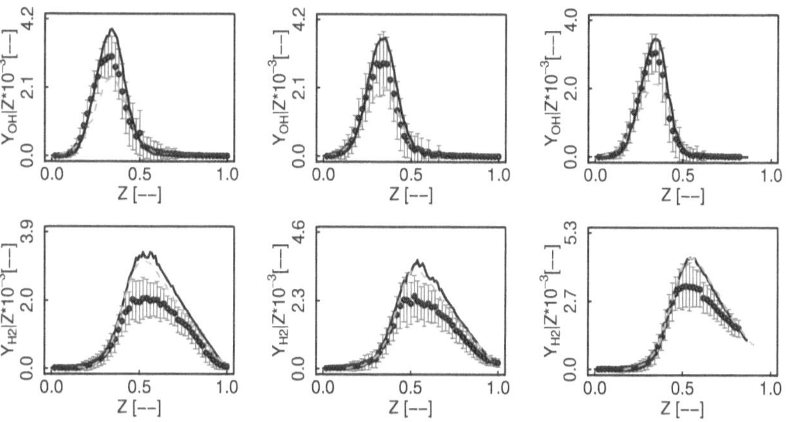

Figure 5.20: Conditional mean mass fraction of OH and H_2 for Flame D. Black line non-premixed FTACLES, green dashed line extended FPV[67], circles the experimental data and error bars RMS.

The non-premixed FTACLES captures better CO_2 behavior in the rich region at $x/d = 7.5$, while the FPV-e model describes more precisely the peak value close to stoichiometry in Figure 5.19. The higher CO_2 concentration comes from a faster fuel consumption, therefore the non-premixed FTACLES predicts higher CO values.

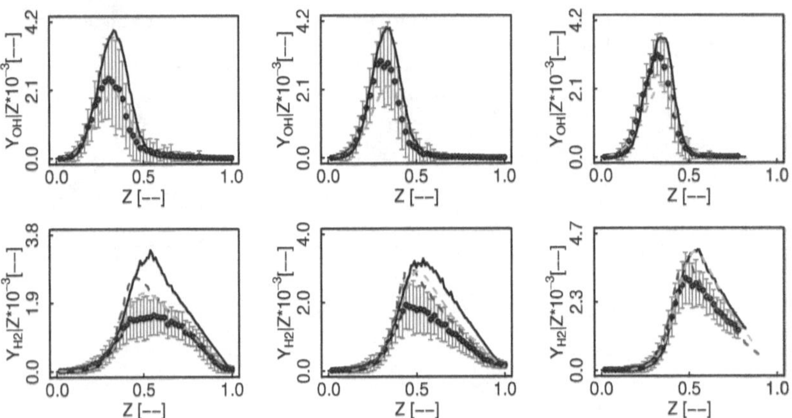

Figure 5.21: Conditional mean mass fraction of OH and H_2 for Flame E. Black line non-premixed FTACLES, green dashed line extended FPV[67], blue dashed line ESF[13], circles the experimental data and error bars RMS.

The non-premixed FTACLES slightly over estimates OH at $x/d = 7.5$ and $x/d = 15$, and it is very good agreement with the experimental data in the last column of Figure 5.20. Hence non-premixed FTACLES outperforms the FPV-e model for the OH predictions, while both of the models converge for H_2. The same OH trend is to be appreciated in Figure 5.21, namely an over prediction of the non-premixed FTACLES and an underestimation of the FPV-e.

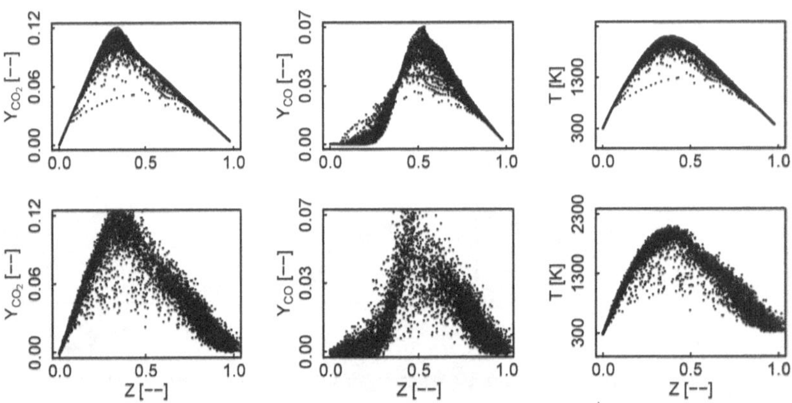

Figure 5.22: Flame D comparison between non-premixed FTACLES(top) and experimental data (bottom) CO_2, CO and temperature scattering at $x/d = 15$.

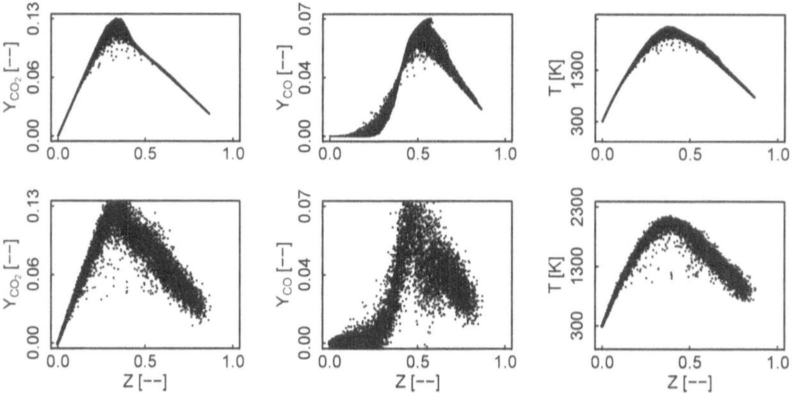

Figure 5.23: Flame D comparison between non-premixed FTACLES(top) and experimental data (bottom) CO_2, CO and temperature scattering at $x/d = 30$.

The good results demonstrate the efficiency of the non-premixed FTACLES strategy for the source term estimation. Within a counterflow flamelet manifold the chemical source term varies by orders of magnitude as function of K. The model exploits the stiffness of the source term, i.e. the abrupt variation in a thin region of the flame, and significantly reduces it through the physical space filtering operation. Figure 5.24 illustrates the benefits of the approach by comparing the effect over the source term with CO_2 transformation.

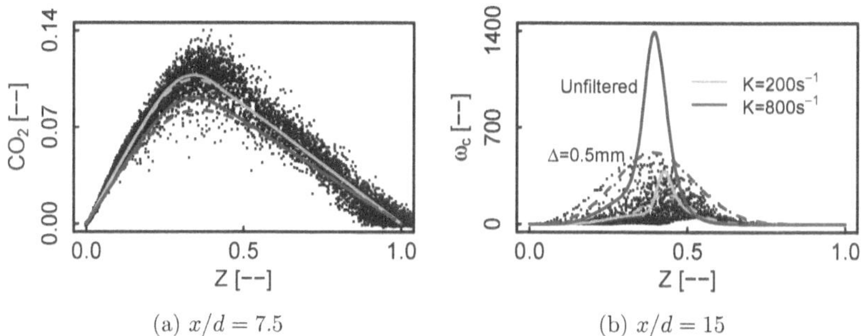

(a) $x/d = 7.5$ (b) $x/d = 15$

Figure 5.24: CO_2 and ω_c scattering for the variable Δ case, together with two flamelets and their profile transformation for $\Delta = 0.5$mm.

Figure 5.24a presents the experimental CO_2 scatter plot along a transverse section, together with two flamelets. The filtering operation causes a negligible impact for $K = 200s^{-1}$, while the peak value diminution for $K = 800s^{-1}$ is of 11%. Figure 5.24b evidences that this diminution, apparently not so significant on CO_2, is considerably higher on the chemical source term. It is reduced to one half for the slower flamelet, while for the faster it corresponds to one third. This feature actually consists in one of the model strengths, as it provides an adequate control of the source term at the same time that the species information is conserved.

5.4   Sensitivity analysis

Relying on the promising results obtained, a sensitivity analysis with respect to the filter size is now carried out for Flame D. The filter size analysis can be decoupled into two aspects: the adequate (Z, c) prediction on one side, and the corresponding retrieved flame structure, e.g. temperature and species, on the other side. The former is linked both to the wrinkling function and to the model, while the second depends uniquely on the formalism itself. The objective is to address the second issue, namely the model response when the flow field and the parameterizing variables remain unaltered. It was shown in Section 3.2 that the flame structure described by a given (Z, c) point on a filtered manifold might significantly change not only as a function of the filter size, but also depending on the specific variable being considered. Thus the analysis can be practically interpreted as the model's ability to deliver the correct values when a coarser mesh is used.

From the practical point of view, setting up the case can be achieved following two different options that may be based on Δ/h ratio:

- In the first one the relation is maintained, thus involving a mesh modification. The main objective of directly filtering one-dimensional flamelets is then to utilize the tabulated chemistry in numerical grids whose cell size is much bigger than that of the detailed chemistry simulations. However, in the context of LES an additional limiting criteria for the cell size is set based on the solution of the energy containing turbulent structures. The $\Delta/h = 1$ approach presents the inconvenience that a significant mesh coarsening could hinder the resolution up to the inertial range, while a refinement might lead to unaffordable mesh size at the same time that it would decrease the filter effect.

- The second option is to consider diverse Δ/h ratios and so the numerical grid is kept unaltered. The flamelet filter does not correspond anymore to the numerical grid cell size h, but it plays the role of a test filter, and the number of mesh points used to resolve the filtered flamelet varies along the domain. The inconvenience of this approach is that a coupling function should be introduced between the numerical grid and the table filter size.

The second approach has been chosen and two-dimensional tables $\varphi = f(Z, c)$ with fixed filter size, namely $\Delta = 2, 3, 4$ mm, were generated instead of three dimensional tables $\varphi = f(Z, c, \Delta)$ as used on the above presented results.

The filter size variation impacts both the model and the wrinkling function performance, hence, this feature has been taken into consideration in the wrinkling function formulation for the condition where the mesh size and the filter size do not coincide. As a result, the mean mixture fraction, velocity and temperature results converge for all the filter sizes, as shown in Figure 5.25, which serves as departure point for the subsequent analysis. Figure 5.26 presents the progress variable behavior, where the mean profiles coincide, while RMS discrepancies are observed for all the sections.

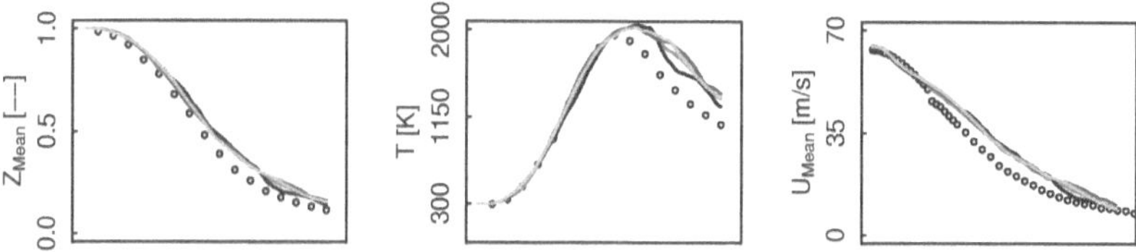

Figure 5.25: Z, T and U mean profiles for cases for different filter sizes. Black line $\Delta = 2$mm, blue line $\Delta = 3$mm, green line $\Delta = 4$mm, red line variable Δ, and circles the experimental data.

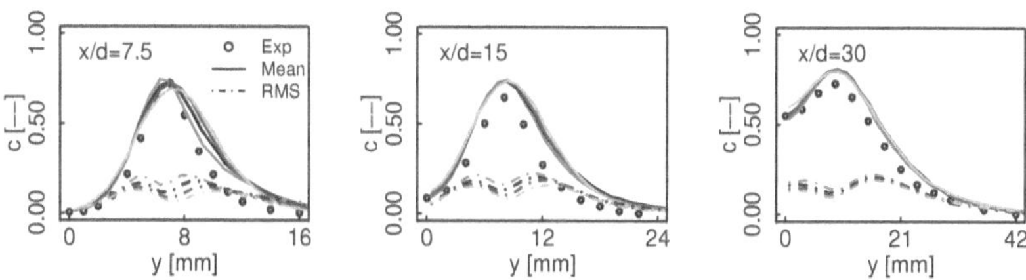

Figure 5.26: Progress variable mean and RMS radial profiles with ξ for different filter sizes. Black line $\Delta = 2$mm, blue line $\Delta = 3$mm, green line $\Delta = 4$mm, red line variable Δ, and circles the experimental data.

5.4.1 Effect of filtering on species prediction

Individual profiles difference

Though mean c profiles coincide, this response is not uniformly transmitted to all the retrieved species. Figure 5.27 depicts the transformations taking place as a result of the filtering operation. At $x/d = 7.5$ a peak value deviation with increasing filter size is observed for CO_2, while for CO a slight profile expansion towards the oxydizer can be recognized. Though OH profile follows c trend, the decrease is considerably higher, so that the peak value significantly falls for the biggest Δ. The slightly non-coincident behavior between CO_2 and CO, as well as OH unequivocal variation as function of Δ hold as well for $x/d = 15$ and $x/d = 30$.

The observed behavior obeys the flamelet displacement along the manifold due to the filtering operation. Two aspects dictate the magnitude of the discrepancy with respect to the original manifold: first the flamelet trajectory deviation, and second the considered species filter sensitivity. The former depends on the profile thickness, i.e. on the relation between δ_c and δ_{Y_i} as already introduced in Section 3.2.3. For instance, T, μ, CO_2 and H_2O possess a quite similar shape and δ_i with respect to c, hence the error is mostly determined by the difference between the original flamelet and the new trajectory, i.e. it depends uniquely on c. For the case of CO, OH and ω_c on the contrary, the filtering operation severely alters their parameterization as function of c, and the considered variable transformation should be additionally taken into account.

The c-Y_i correlation variation as a function of the profile thickness has fundamental implications, as it determines the behavior of each species mean and fluctuations.

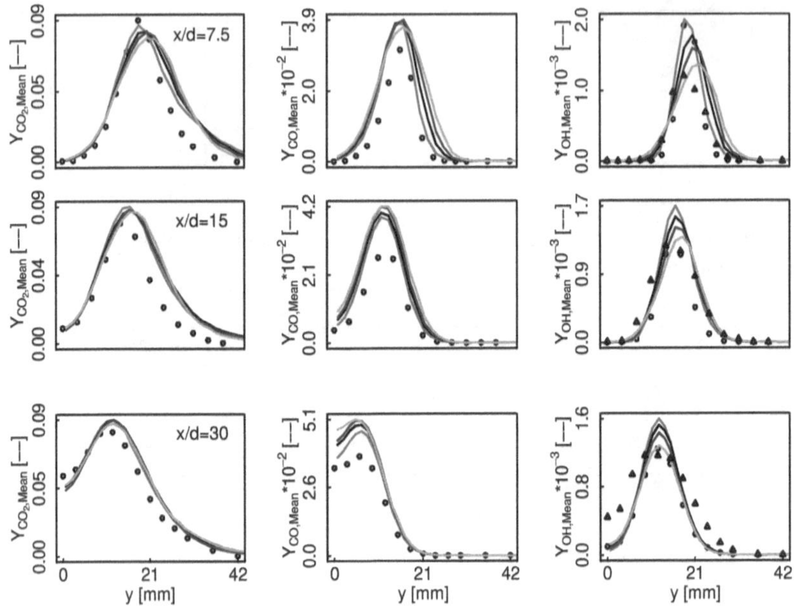

Figure 5.27: CO_2, CO and OH profile transformation due to the filtering effect. Black line $\Delta = 2$mm, blue line $\Delta = 3$mm, green color $\Delta = 4$mm, and circles the experimental data.

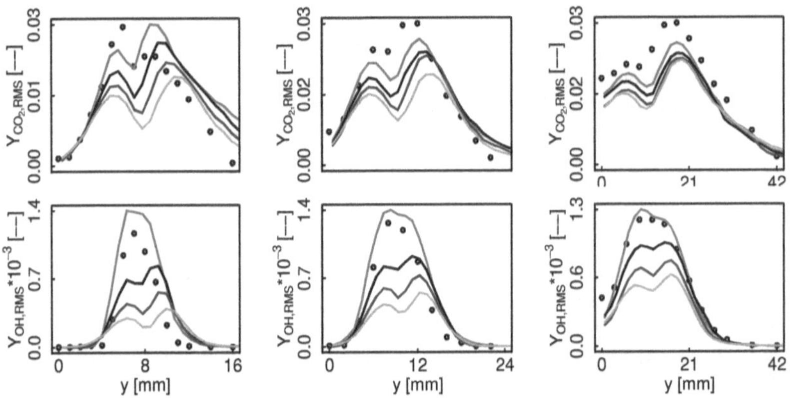

Figure 5.28: CO_2 and OH RMS radial profiles. Black line $\Delta = 2$mm, blue line $\Delta = 3$mm, green line $\Delta = 4$mm, red line variable Δ, and circles the experimental data.

The case of an unequivocal relation between c and an arbitrary Y_i for any given Δ coincides with a conditional averaging approach. It follows that $Y_{i,Mean}$ can be determined by c_{Mean} and c_{RMS} can be to a good extent transposed to $Y_{i,RMS}$. If this is not the case, not only the mean but as well the species fluctuations will present a modified behavior, and should be individually considered as shown in Figure 5.28. While CO_2 follows the same trends as c, for OH the deviation with respect to the three-dimensional table substantially increases even altering the curve shape.

Species scattering

Figure 5.29 shows a snapshot of the accessed CO_2, CO and OH as function of c for the different Δ. The progress variable fluctuations decrease with the filter size, which explains the maximum c decrease even though the right c_{Mean} is retrieved. CO_2 linearly evolves as function of the progress variable and the trend holds throughout the different filter sizes. Though CO is less correlated, it does have a similar thickness, and therefore the filtering operation does not significantly modify its profile. For OH

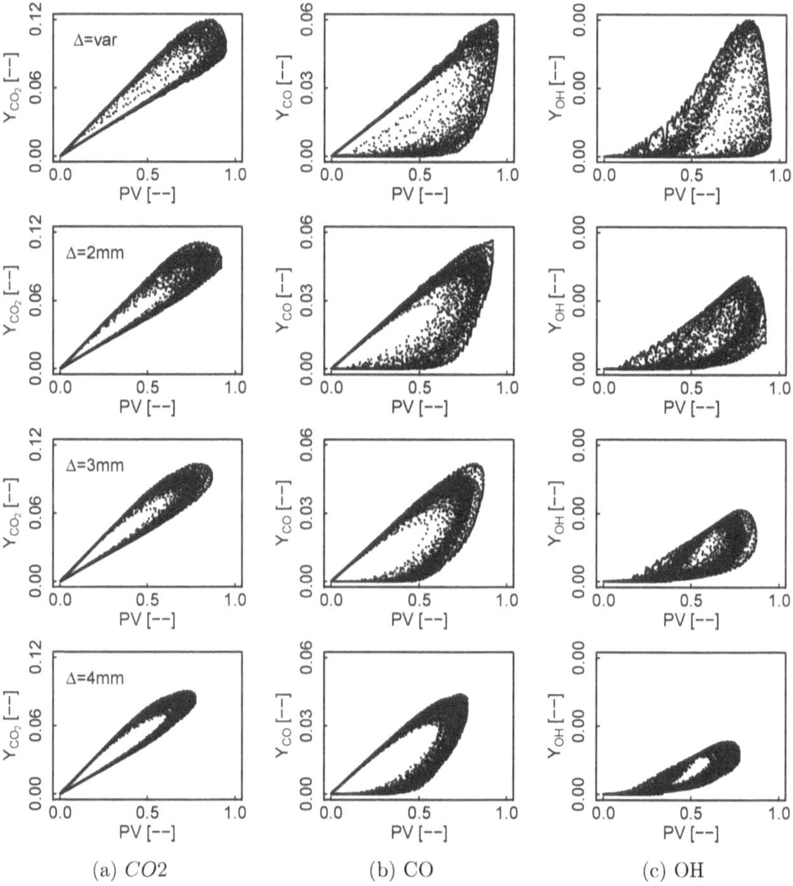

(a) CO2   (b) CO   (c) OH

Figure 5.29: Manifold transformation due to the filtering operation for Y_{CO_2}, Y_{CO} and Y_{OH} as a function of c at $x/d = 7.5$.

on the contrary, the filtering operation eliminates much of the unfiltered profile features. This linearization of c-OH relation results into predictions that considerably sink for high c.

The results are encouraging as they indicate that reasonably good mean species predictions can be obtained even for non linear relations between c and a given Y_i. The response is conditioned by δ_{Y_i}, thus an adequate estimation will be obtained if it lies in the same order of magnitude as δ_c, while the deviations will increase for thinner profiles. The prediction of the species fluctuations on the other side appears to be affected to a higher degree, due to a combined effect of c fluctuations decrease and Y_i transformation. Moreover, the correct Z and c profiles computation is a necessary but not sufficient condition for the retrieved variables conservation. For instance, it can be unequivocally stated that for $\Delta = 4$mm, it results impossible to adequately reproduce OH results, regardless the agreement in the c prediction.

5.4.2   Effect of filtering and modified turbulence interaction

There are two fundamental phenomena taking place as Δ increases, the first corresponds to the modification in the species relation to c, as already addressed above, the second corresponds to a modified turbulence interaction. Figure 5.26 pointed out an uncorrelation between c_{Mean} and c_{RMS}, where the appropriate prediction of the former does not ensure the right estimation of the latter. The RMS values decreased with the filter size, at the same time that c_{Mean} remained unmodified. Figure 5.30 shows the accessed manifold region for three snapshots, aiming to carry out a qualitative description, certainly taking into account instantaneous data limitations.

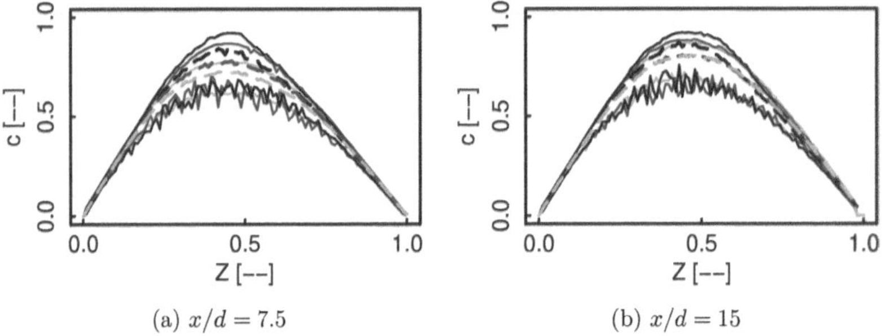

(a) $x/d = 7.5$ (b) $x/d = 15$

Figure 5.30: Snapshot of accessed c vs Z and the conditional mean, for two different transverse sections and filter sizes. Solid lines the manifold limits, dashed line conditional mean. Black line $\Delta = 2$mm, blue line $\Delta = 3$mm, and green line $\Delta = 4$mm.

There is a clear inverse relation between Δ and the top limit, so that $\Delta = 2$mm always reaches the highest values. This behavior is not observed in the lower region, where despite the high oscillations the three filter sizes are located within the same range. Therefore the amplitude decrease with Δ obeys the top limit modification and is not affected by the lower limit.

The conditional mean is much closer to the top than to the bottom limit, which owes to the big dispersion, i.e. low point density, in the lowest part of the manifold. These results suggest that the observed fluctuations decrease comes from an underestimation of the values lying above and not under the mean. Since the highest part of the manifold consists of low strained flamelets, they are less sensitive to the filtering operation due to their big thickness. Acknowledging that the accessed lower part of the manifold remains relatively constant, points out strong transformations in these higher K regions as playing an important role in the top limit under prediction.

Fluctuations modification

Both the filtering operation and the correction exerted by ξ contribute in an intricate manner to the fluctuations decrease. The sensitivity analysis test has been designed so that the flow field and the parameterizing variables remain unaltered. It is worth noting that this condition holds for the mean values, however it does not hinder that dissimilarities appear on the corresponding fluctuations as has already been pointed out in Figure 5.26 for the progress variable, and is presented for the mixture fraction and the velocity in Figure 5.31. For instance the model systematically under estimates Z_{RMS}, while the U_{RMS} profiles appear to be less sensitive.

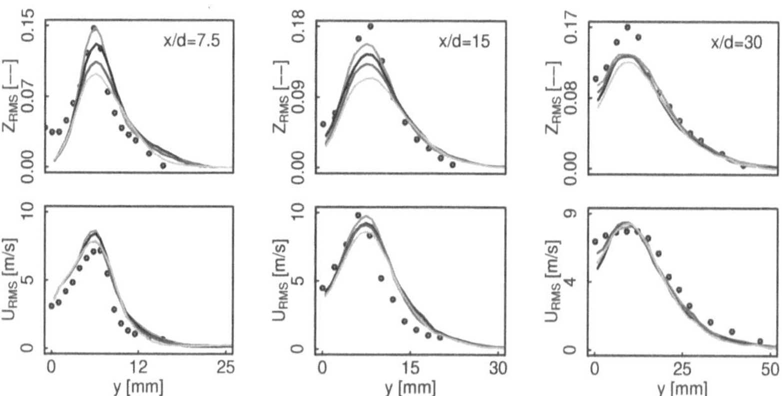

Figure 5.31: Z and U RMS radial profiles for different transverse sections. Black line $\Delta = 2$mm, blue line $\Delta = 3$mm, green line $\Delta = 4$mm, red line variable Δ, and circles the experimental data.

The velocity equation is only indirectly affected by the SGS wrinkling, namely through the progress variable computation and therefore the retrieved molecular viscosity. For Z equation on the contrary, additional to c effect, ξ explicitly increases the diffusivity. The mixture fraction fluctuations filter dependence highlights the relevance of this operation.

Figure 5.32 shows the corresponding μ and μ_{SGS} profiles. Due to the linear relationship between μ and c, the good c_{Mean} agreement results into a small change on the related mean values, namely μ and therefore μ_{SGS}. On the contrary, the filter effect can be unambiguously identified on μ_{RMS}, which undergoes a reduction with increasing filter size. Altogether this has a barely recognizable impact over U_{RMS}.

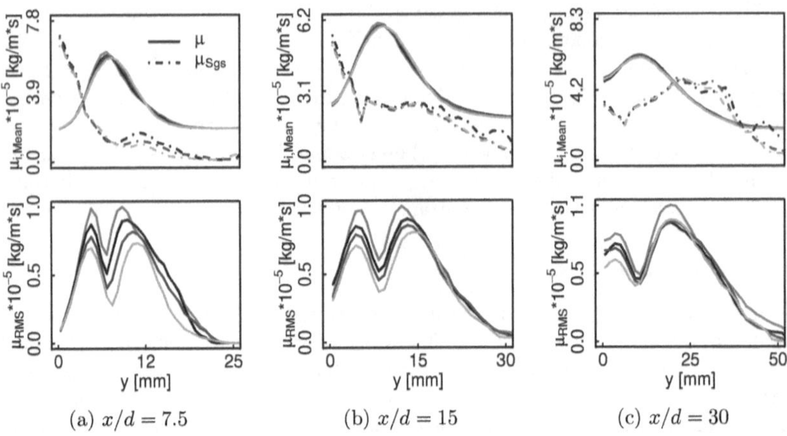

(a) $x/d = 7.5$ (b) $x/d = 15$ (c) $x/d = 30$

Figure 5.32: Molecular and SGS viscosity radial profiles for different filter sizes. Black line $\Delta = 2$mm, blue line $\Delta = 3$mm, green line $\Delta = 4$mm, and red line variable Δ reference.

Figure 5.33 illustrates the combined effect of the filtering operation and the modified turbulence interaction on $\omega_{c,Tot} = \xi \cdot \omega_c$ and $D_{Tot} = \xi \cdot D$. At the first section ξ presents two clear peaks at both of the pilot stream shear layers, which significantly alter D_{Tot} profile, while they do not fully coincide with $\omega_{c,Tot}$. The filtering operation considerably diminishes the latter, e.g. to almost one half between the variable Δ and $\Delta = 4$mm, despite ξ contribution. Downstream the source term difference between the filter sizes decreases, and a profile expansion towards the axis can be appreciated. The smoother ξ radial evolution conserves D_{Tot} shape, besides the amplification. Acknowledging the passive scalar nature of Z, i.e. ξ will not contribute by means of a source term, D_{Tot} increase acts as a numerical diffusion, which explains the fluctuations decrease with increasing filter size, as a consequence of the higher ξ correction.

For the case of the progress variable, additional to D_{Tot}, the filtering operation further contributes to the fluctuations decrease. Considering a filtered manifold, the iso Z displacement $\Delta_c = |c_1 - c_0|$, s.t. $c_1 < c_0$, delivers an $\omega_{c,1}$ which decreases with the filter size, and the ratio $\omega_{c,0}/\omega_{c,1}$ does not necessarily remain constant as Δ changes. Assuming that c_1 corresponds to the same flamelet, it is evident that an increasing filter size will further reduce $\omega_{c,1}$. However, there is another consideration that should be made, namely the manifold transformation due to the filtering operation. As the filter size increases, any Δ_c located in the steady flamelets region of the unfiltered manifold will describe a much more homogeneous set of conditions, as the higher K trajectories undergo a bigger transformation and therefore collapse in the lower part of the manifold. Since the source term increases with K, this homogenization intrinsically reduces $\omega_{c,1}$. Proceeding with the iso Z hypothesis, and assuming that the flow field does not change, it results that the following c displacement $\Delta_c = c_2 - c_1$ will decrease with the filter size, and consequently so will the fluctuations. This agrees with Figure 5.30, as there the lower limit c_1 holds, but c_2 decreases with increasing Delta.

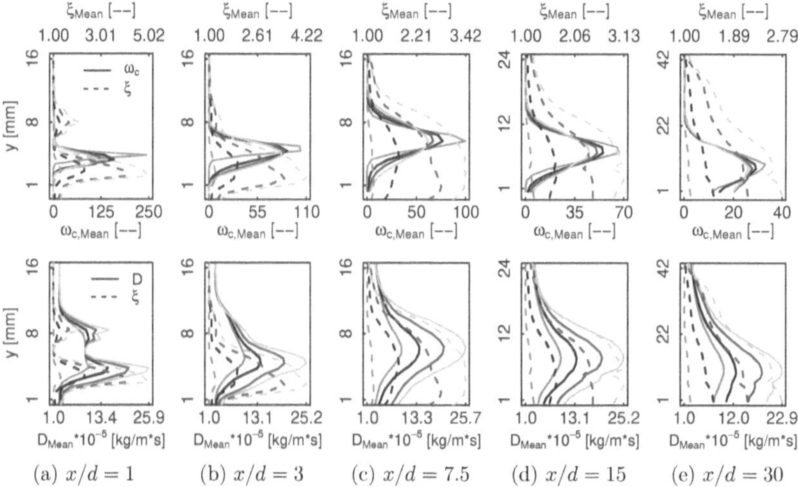

Figure 5.33: Progress variable chemical source term, efficiency factor and total diffusivity radial profiles for different filter sizes. Black line $\Delta = 2$mm, blue line $\Delta = 3$mm, green line $\Delta = 4$mm, and red line variable Δ.

5.4.3 Counterflow flame behavior

In the non-premixed FTACLES contrary to the premixed counterpart, correction source terms are added not only to c but also to Z. Recalling Equation 4.1, these terms are activated based on a comparison between the flamelet and the simulation mixture fraction gradients, as it has been shown that respecting the counterflow conditions is fundamental to avoid non-physical results. The LES resolved parameters (Z, c) used to access the reference ∇Z_{fl} from the manifold correspond to the spatial average within each computational cell. Assuming that the flame front behaves laminar, the FTACLES approach exploits the fact that the profile transformation can only result from the filtering operation. However, if flame front deformation takes place at the SGS level, these parameters depict the combination of higher c surfaces surrounded by lower c regions within the cell, and this has deep implications for the application of the non-premixed FTACLES formalism. The sensor activation no longer indicates the expansion of a laminar counterflow flamelet, since the departure points do not correspond to such a condition, but it points out to a double flame structure transformation, from a subgrid wrinkled flame to an analogous thickened counterflow laminar flame. Chapter 4 presented a thorough analysis of the deviations that multidimensional effects induce between the two gradients. This section considers a second reason for the gradients to dissent, namely SGS wrinkling.

Recalling the enormous relevance of geometrical features on S activation, a convenient starting point consists to identify the domain regions with preferentially radial Z evolution. The consideration of instantaneous K profiles on Figure 5.34 suggests a domain division based on the jet core presence. This feature can be recognized by the centerline constant strain rate at the first two transverse sections, i.e. $\nabla Z = 0$, and consequently a deactivated sensor. The fuel-pilot interface is characterized by a steep K decrease towards a considerably low value at the pilot stream, and this frequently coincides with an active S. The jet core enhances Z evolution in the radial direction and so it promotes S activation. Nonetheless, the discontinuity in S profile demonstrates that solely this condition does not ensure the unequivocal fulfillment of the counterflow flamelet hypothesis, and it points towards the SGS influence. The jet core vanishes at $x/d = 7.5$ modifying the flame front topology in behalf of axial direction gradients which are now added to the radial ones, being this a flame feature, and not a filtering operation effect. The K profile spreads over a broader zone, and it does not follow any clear correlation with respect to S. Further downstream the sensor activation continues to decrease, till it is almost completely turned off at $x/d = 30$.

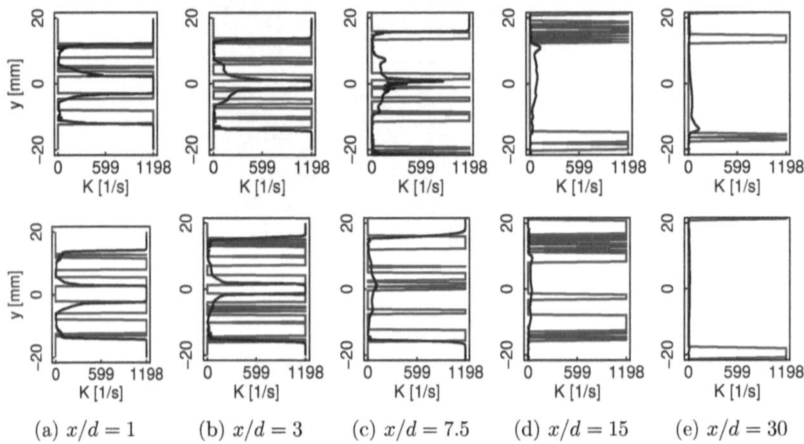

(a) $x/d = 1$ (b) $x/d = 3$ (c) $x/d = 7.5$ (d) $x/d = 15$ (e) $x/d = 30$

Figure 5.34: Instantaneous K radial profiles for $\Delta = 2$mm and 4 mm, top and bottom rows respectively. Black line strain rate, blue line flame sensor.

Expressing these results in Z space Figure 5.35 shows that for the first two transverse sections the K profile is unambiguously split around $Z_{Pilot} < Z_{st}$. The sensor is active on the fuel side extreme, presents a varying behavior within the flame and is activated once again towards the oxidizer side. Thus, apart from the already identified jet core influence, these observations bring to light the presence of a pilot core which was indeed not evident from Figure 5.34. On the rich side K increases in direct proportion to the distance from Z_{Pilot}, as expected from the pure mixing jet core condition at $Z = 1$, while on the lean counterpart K remains rather at a constant low value. This clear division at Z_{Pilot} as well as the rising behavior towards $Z = 1$ disappears at $x/d = 7.5$, indicating the vanishing of the boundary conditions influence.

Figure 5.36 presents ∇Z_{sim}, ∇Z_{fl}, and S mean profiles along the first two transverse sections. There is a directly increasing S activity with the filter size for the two-dimensional cases. For instance at $x/d = 1$, the sensor tends towards a constant activation at the fuel side extreme for $\Delta = 4$mm, non-conforming events increase as Δ decreases, and finally the counterflow condition is just seldom satisfied for variable Δ. The observation of the composing elements of Equation 4.1 indicates that the behavior is controlled by the reference term, i.e. ∇Z_{fl}. This is a result of the flamelet redistribution along the manifold, which causes the mixture fraction gradient in a fixed point (Z, c) to decrease as Δ rises. The decreasing ∇Z_{sim} is coherent with the thickening procedure, and in comparison with the reference, it presents a much more homogeneous behavior among the considered cases.

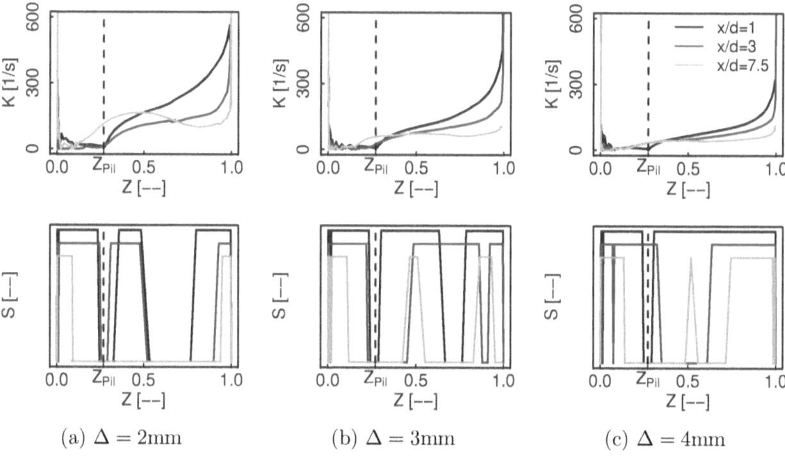

(a) $\Delta = 2$mm   (b) $\Delta = 3$mm   (c) $\Delta = 4$mm

Figure 5.35: Instantaneous K and S radial profiles for three different transverse sections and three filter sizes in Z space.

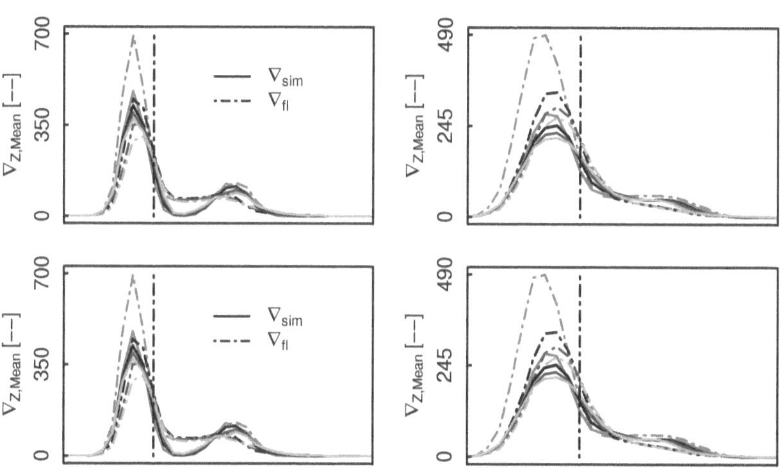

Figure 5.36: ∇Z and S profiles for cases $S3$ at $x/d = 1$(top) and $x/d = 3$(bottom). Black line $\Delta = 2$mm, blue line $\Delta = 3$mm, green line $\Delta = 4$mm, and red line variable Δ.

The dominance of ∇Z_{fl} on S activation, together with the much moderate ∇Z_{sim} variation indicate that though the numerical solution slightly varies with Δ, its physical interpretation substantially changes. The coincidence between the flamelet and the simulated thickness, i.e. an active S, reveals an homogenization of the computational cell. Hence the modification takes place at the SGS level, the wrinkled flame is substituted by a thick laminar flame, and as long as the jet core exists it might strongly resemble a counterflow flame.

Figure 5.37 depicts $\omega_{c,i}$ terms evolution in Z space. Recalling the observation made on Figure 5.35, the sensor appears to be active at both of the extremes, while the inner region of the flame front presents an intermittent behavior. The high K values found in the fuel-pilot interface result into correction terms in the same order of magnitude, but opposite sign with respect to the filtered chemical source term Therefore, close to the fuel side the model correction terms play a fundamental role, as they compensate the chemical source term expansion resulting from the filtering operation. The low K values in the oxidizer side explain the almost negligible correction terms, and thus the lack of model effect despite an active sensor. The behavior observed on the top right figure, i.e $x/d = 1$ and $\Delta = 4$mm, is particularly useful as it illustrates the already mentioned comprehensive flame structure transformation. The continuous S activation, covering the chemical source term vast majority, indicates that the resolved structure of the original wrinkled flame can be effectively described employing an analogous laminar counterflow flamelet when subjected to a filtering operation with $\Delta = 4$mm.

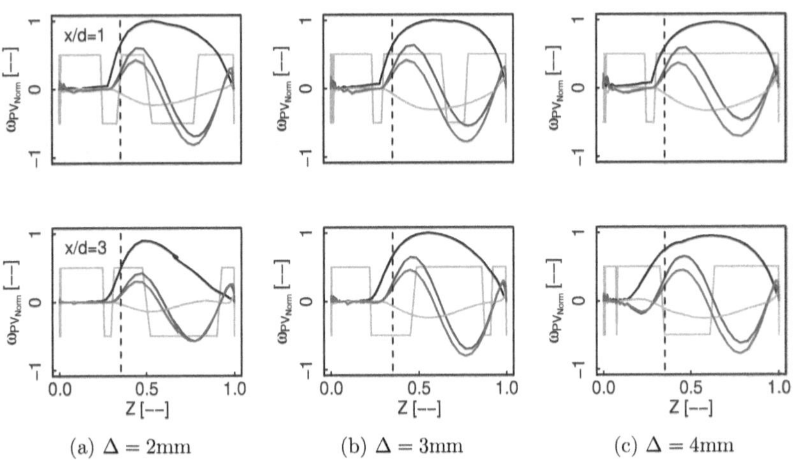

Figure 5.37: Instantaneous normalized ω_{PV} radial profiles for $\Delta = 2,3,4$ mm. Black line chemical source term, blue line diffusion correction, green line convection correction, purple line total correction, light gray line flame sensor, and the dashed line is c_{max} location.

The almost negligible S activation for the variable Δ case highlights a well known fact, namely that the flame structure described by a wrinkled (Z, c) pair significantly differs from that of the analogous laminar flamelet. This is not surprising, and it is

for this reason that further considerations, e.g. Z_v^2 on a β-PDF approach, are taken
in order to correlate the resolved (Z, c) with a laminar flamelet manifold. However,
it results useful to spend a few words on the FTACLES performance under such
conditions. Figure 5.38 presents the source terms behavior for the variable Δ case,
and the first aspect that calls the attention is that correction terms are significantly
smaller than the scenario presented by the two-dimensional tables. The resolved
(Z, c) point corresponds to a flamelet whose thickness lies in the same range than
the filter. Thus Z and c induced profile deviations are small, as it can be seen
from the convective and the diffusive contributions. The chemical source term, on
the other side, has a significantly lower profile thickness, and therefore will undergo
a significant transformation as already exemplified in Figure 5.24. It can then be
said that under those conditions, the flame structure transformation due to the
filtering operation will be mainly addressed by the wrinkling function, while the
model correction terms are omitted.

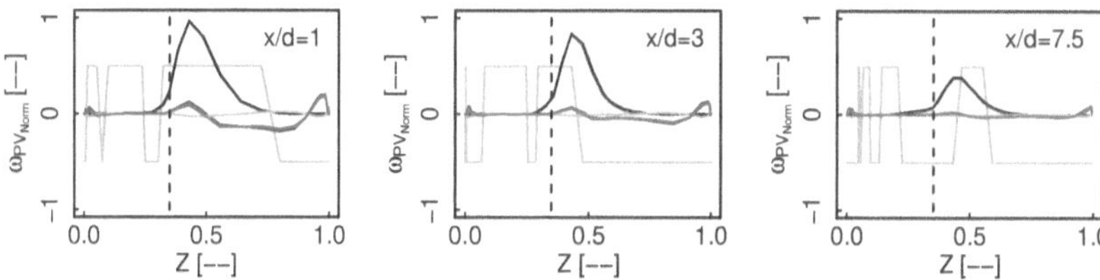

Figure 5.38: Instantaneous normalized ω_{PV} radial profiles for a three-dimensional
table at different radial sections. Black line chemical source term, blue line diffusion
correction, green line convection correction, purple line total correction, light gray
line flame sensor, and the dashed line is c_{max} location.

5.5 Perspectives: pollutant formation estimation

Adequate pollutant estimation is a very important feature to be considered in the
development of combustion models. This section presents a succinct view of the
way NO estimation could be incorporated in the current formalism. The first part
illustrates the problem that arises if NO is directly estimated from a steady flamelet
table. Subsequently some key issues specifically related to the non-premixed filtered tabulated chemistry approach are cited and guidelines for the further work are
proposed.

5.5.1 Direct NO estimation from steady flamelet table

Two major approaches exist for NO estimation in flamelet-based combustion modeling: the direct table retrieval and the solution of an additional transport equation.
Though simpler to implement, the former method tends to highly over estimate
NO, the reason for this behavior has been widely discussed in literature and will be
briefly summarized here. Steady flamelet methods employ the high Da assumption,
i.e. the chemical timescale is significantly smaller than the flow timescale, therefore
the flame structure can immediately respond to fluctuations in the flow field and

can be described by a steady state condition. NO formation is a slow process, with a timescale largely exceeding that of the major species and often greater than the considered device residence time. It follows that while the flame structure at an arbitrary flame location might be adequately described when the flamelet approach is employed, the NO concentration will generally be lower than that predicted by steady flamelets.

Figure 5.39 compares non-premixed FTACLES and FPV mean NO profiles for Flame-D. The instantaneous values have been directly retrieved from the look-up table, i.e. $NO = f(Z, c, \Delta)$ and $NO = f(Z, Z_v^2, c)$ respectively, following the same procedure for the species estimation in Section 5.3.2. Both of the models largely over estimate the concentrations by a factor that ranges between $4x$ and $5x$. The results coincide with those corresponding to the *NO from library*, i.e. the direct table retrieval, in[66]. This study analyzed NO production for Flame-D with the FPV method coupled to different pollutant formation modeling strategies, and demonstrated the unavoidable need to employ a NO transport equation.

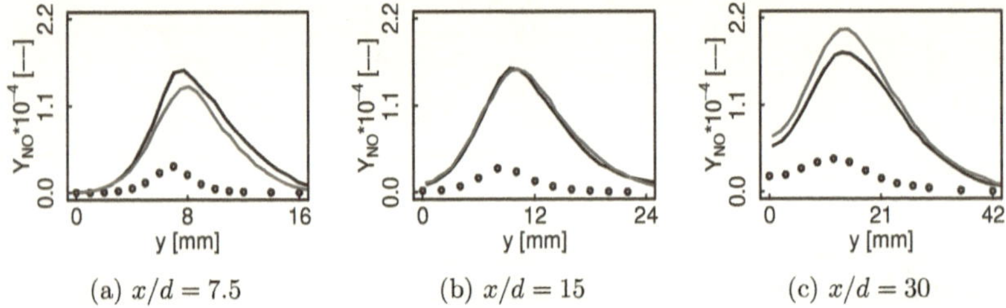

(a) $x/d = 7.5$ (b) $x/d = 15$ (c) $x/d = 30$

Figure 5.39: Comparison of non-premixed FTACLES and FPV NO mean values for Flame D. Black line non-premixed FTACLES, purple line FPV and circles experimental data.

The similarity between the non-premixed FTACLES and FPV results is coherent with the previous species predictions. The pollutant prediction deviation is linked to the steady state assumption for the manifold generation, and does not depend to a first level neither on the flame front resolution nor on the SGS modeling as the overall effect of the coupled filtering and wrinkling function is equivalent to the SGS modeling performed by the presumed β-PDF. This is a preliminary encouraging feature as it suggests that NO prediction employing a transport equation might significantly improve the results for the non-premixed filtered tabulated chemistry model.

5.5.2   Perspectives

The NO computation directly from the look-up table significantly over predicted the experimental values. On the contrary solving a transport equation for NO is a well known pollutant formation estimation strategy, where the tabulated chemistry coupling takes place through the chemical source term. This approach has been widely employed for flamelet-based approaches, e.g. FPV[66] and FGM [148][42], and a substantial improvement with respect to the direct table retrieval has been reported. Apart from the above mentioned inconveniences of using direct table retrieval for steady flamelet techniques, there are two additional difficulties particular to the filtering operation:

- Manifold modification: The filtering operation displaces the flamelet profiles downwards within the manifold, i.e. they are shifted in the direction of the pure mixing line, the filtering sensitivity being a function of the strain rate. This leads to the manifold homogenization discussed in Chapter 3, where the lower strained flamelets cover larger manifold regions as the filter size increases. NO is highly sensitive to the flamelet strain rate, and this filtered flamelet redistribution might further contribute to NO over estimation.

- Profile trimming: as it has been widely discussed, the filtering operation reduces the peak values for non-monotonic variables, as is the case for NO and $\dot{\omega}_{NO}$. Not only the value itself will correspond to a steady state condition, which in general does not hold, but this value is further modified by the filtering operation. Due to the low profile thickness that characterizes the chemical source terms, they are particularly sensitive and the effect of its decrease should be taken into consideration.

For the above explained reasons, further research should be carried out to appraise the performance of the NO transport equation approach in the current formalism. Concretely, studies can be performed to investigate the relevance of the following features:

- Adiabatic assumption: It is well known that NO formation is highly dependent on temperature. So far the non-premixed filtered tabulated chemistry approach employs a 3-dimensional parameterization $\varphi = f(Z, c, \Delta)$. As a first step the pollutant prediction can be done in the adiabatic condition. Then the model sensitivity to the addition of h as a fourth dimension to take into account radiation effect should be appraised.

- Transport equation closure: NO computation by means of a conservation equation introduces the challenge of the unclosed terms characteristic from the LES filtering. A consistent strategy should be proposed in order to link the filtered $\dot{\omega}_{NO}$ with the transport equation.

- Progress variable definition: for the case of premixed combustion since NO formation might continue after a given c definition has reached unity, extended c definitions including Y_{NO} have been proposed in order to improve the quality in $\dot{\omega}_{NO}$ resolution. Though the phenomenology in non-premixed flames is completely different, and NO formation takes place in the reacting region, i.e. close to Z_{st}, the influence of c definition should be assessed. In this scenario the consideration of NO in c definition might be justified considering its variation with K and therefore the progress variable's capability to adequately resolve this behavior.

5.6 Conclusions

The chapter has presented the first evaluation of the non-premixed FTACLES model on a turbulent flame. The results demonstrated that the formalism coupling with a SGS modeling function can adequately describe a wrinkled flame front condition. The predictions for both the major stable species and the minor ones accurately correspond with the undergoing physics. Putting them into context, the obtained results have a deep theoretical implication as they confirm the idea that SGS closure in diffusive combustion can be derived based on filtering arguments, and not only based on statistical approaches.

Section 5.3 extensively appraised the non-premixed FTACLES model performance altogether with own FPV results and the extended FPV taken from [67]. The objective of this comparison was a clear separation between the turbulence combustion model and the chemical database influence. For instance, the conditional averaging overestimation for CO and H_2 while CO_2 is satisfactorily retrieved in Figures 5.18 - 5.21 could misleadingly point out to a profile distortion and an error introduced due to the non-premixed FTACLES model. On the contrary, the striking coincidence with the FPV results indicates that to a big extent the error is related to the steady state counterflow flamelets employed for the chemical database generation. With respect to the use of a presumed PDF the non-premixed FTACLES presents the advantage that no assumption is done concerning the statistical distribution, though for a non-laminar flame front at the SGS level the modeling component is shifted to the wrinkling function. Furthermore, the methodology directly introduces the effect of the numerical grid size into the combustion model, being thus highly consistent with the LES technique.

The comparison between the premixed and non-premixed FTACLES showed that in the absence of SGS modeling, both approaches deliver analogous results. This is a fundamental outcome with two corollaries: first, it justifies the extension of a wrinkling function from the premixed to the non-premixed condition. The system response proves to be controlled by the filter size, and the contribution of each the model closure terms plays a secondary role. Second, it reinforces the hypothesis that if a SGS function is included, the non-premixed formalism should deliver similar results as the premixed counterpart.

Relying on the promising model results, a sensitivity analysis was carried out in terms of the filter size. The same numerical grid was employed and two-dimensional tables $\varphi = f(Z, c)$ with fixed Δ were generated. Besides the expected correct predictions for species linearly correlated with the progress variable, good results were obtained for non linearly correlated species. The response appeared to be conditioned by δ_{Y_i}/δ_c ratio. An adequate estimation was obtained for values close to unity, while the deviations increased for smaller ratios. This is a positive feature as it suggests that the method is able to correctly predict major species even in significantly coarse meshes. Nevertheless, for species with considerably lower thickness than the progress variable, the formalism inevitably introduces an error due to a profile linearization, and this effect increases with the filter size. From the practical point of view this implies that there is minimum grid resolution to be respected, for the model to adequately predict a given species.

The sensitivity analysis identified a decrease both in c and Z fluctuations with the filter size. For the mixture fraction this diminution obeys the total diffusivity modification on behalf of ξ, which finally acts as a numerical diffusion. The filtering operation substantially modifies the chemical source term distribution along the manifold due to the flamelet displacement and to profile trimming. Hence, the progress variable reduction obeys both the filtering operation and the modified diffusivity effect.

The sensor response appears to be consistent with the decreased flame front deformation. The computational cell SGS scalar distribution undergoes an homogenization with increasing filter size. Thus, the wrinkled flame is substituted by an analogous laminar flame with a considerable higher thickness.

This chapter evidenced the enormous potential of the non-premixed FTACLES method. The model performed significantly well employing a three-dimensional tabulation strategy, where the numerical grid was coupled with the model by the third parameter, i.e. the computational cell size. Moreover, a secondary test, with a test-filter-like approach and two dimensional parameterization in terms of Z and c, showed a good model response for coarse numerical grids.

Chapter 6

Summary and outlook

This work addresses the application of non-premixed filtered tabulated chemistry as a turbulent combustion modeling strategy in the LES framework. For this, a systematic approach is followed where first the numerical implementation is verified, afterwards validation is done on a coflow laminar diffusion flame, and finally the model full capabilities are assessed on the more complex turbulence condition. The promising results obtained on the coflow laminar diffusion flame encouraged the further model appraisal on a more complex turbulent configuration. On the turbulent configuration the formalism coupling with a SGS modeling function adequately described a wrinkled flame front condition. This work evidences the enormous potential of the non-premixed FTACLES method as it confirms the idea that SGS closure in diffusive combustion can be derived based on filtering arguments, and not only based on statistical approaches.

The effects of the filtering operation on non-premixed flamelets have been thoroughly appraised. The flame structure modification has been described considering an individual flamelet and the entire manifold. It turns out that the filter effect varies considerably along the manifold as function of K due to the flame thickness reduction at higher strain rates. The filter effect also differs for each variable within a flamelet, depending on the profile shape and thickness. As a result, there might be a non linear relation between Z, c and the retrieved species φ under the filtering operation. Non-monotonic variables are substantially modified due to the trimming of the peak value. Considering the progress variable, this transformation shifts the flamelets downwards within the manifold. The deviation in the retrieved variables prediction increases with the filter size. Two aspects dictate the magnitude of the discrepancy: first the significance of the flamelets deviation within the manifold, and second the considered species filter sensitivity.

The different filtering effect between a premixed and a non-premixed manifold has been addressed. The relation between $Z - c$ is altered on non-premixed flames, thus the species behavior in chemical space is inevitably modified with increasing filter size. For the premixed condition on the contrary, c transformation impacts exclusively the physical space. It follows that the flame structure transformation in chemical space depends on the correlation between the considered species and the progress variable. For highly correlated variables, e.g. T, the filtered manifold strongly resembles the unfiltered one independent of the filter size, hence decoupling the species prediction from the flame thickness. This is a fundamental issue as it completely modifies the interpretation of a filtered flame on the two approaches.

The performance of the non-premixed filtered tabulated chemistry formalism has been assessed on a laminar coflow diffusion flame. The model appears to be considerably sensitive to the flame dimensionality. Unphysical profile distortions arise when points described by highly strained regions of the manifold, and consequently having big model correction terms, do not comply with the one-dimensional flamelet hypothesis. A flame sensor based on the mixture fraction gradient, with a tolerance to take into account the numerical grid resolution, has been introduced. Simulations performed using the model coupled with the flame sensor have been compared against the direct filtering of a reference solution, and good agreement has been obtained for the laminar flame. The sensor determined model activation adequately represents the underlying physics behind flame filtering and so it endorses the consistency of the numerical procedure as a thickening operation.

Expanding on the laminar coflow diffusion flame findings, using three different filter sizes, it turns out that: Far from the centerline the flame front does not only satisfy the counterflow hypothesis, but K remains unaltered after the filtering operation, what can be understood as the filtering of the same trajectory, namely of the same flamelet. The filter effect at the radial reaction zone is then transported throughout the domain and is the cause for the centerline profile modifications, where the model is not active. The profile extension due to the filtering process is mainly driven by the model correction terms, while the decrease in the peak values of non monotonically evolving variables depends on the filtered parameters that enter the transport equation. The sensor adequately identifies the multidimensional effect at the centerline, and its active range both in Z as in physical space changes throughout the domain, in accordance with the strain rate decrease and higher flame thickness.

This work presents the first evaluation of the non-premixed FTACLES model on a turbulent flame. As application object, the Sandia flames D and E have been selected, as they are well documented in terms of experimental data and model validation in the literature. The results demonstrate that the formalism coupling with a SGS modeling function can adequately describe a wrinkled flame front condition. The model performs significantly well employing a three-dimensional tabulation strategy, where the numerical grid is coupled with the model by the third parameter, i.e. the computational cell size. The predictions for both the major stable species and the minor ones accurately correspond with the undergoing physics. Putting them into context, the obtained results have a deep theoretical implication as they confirm the idea that SGS closure in diffusive combustion can be derived based on filtering arguments, and not only based on statistical approaches.

The comparison between the premixed and non-premixed FTACLES shows that in the absence of SGS modeling, both approaches deliver analogous results. This is a fundamental outcome with two corollaries: first, it justifies the extension of a wrinkling function from the premixed to the non-premixed condition. The system response proves to be controlled by the filter size, and the contribution of each the model closure terms plays a secondary role. Second, it reinforces the hypothesis that if a SGS function is included, the non-premixed formalism should deliver similar results as the premixed counterpart.

A sensitivity analysis has been carried out in terms of the filter size for Flame D. The same numerical grid has been employed and two-dimensional tables $\varphi = f(Z, c)$ with fixed Δ have been generated. Besides the expected correct predictions for species linearly correlated with the progress variable, good results have been obtained for non linearly correlated species. The response appears to be conditioned by δ_{Y_i}/δ_c ratio. An adequate estimation has been obtained for values close to unity, while the deviations increased for smaller ratios. This is a positive feature as it suggests that the method is able to correctly predict major species even in significantly coarse meshes. Nevertheless, for species with considerably lower thickness than the progress variable, the formalism inevitably introduces an error due to a profile linearization, and this effect increases with the filter size. From the practical point of view this implies that there is minimum grid resolution to be respected, for the model to adequately predict a given species. Thus, the secondary test, with a test-filter-like approach and two dimensional parameterization in terms of Z and c, shows a good model response for coarse numerical grids.

This work has evidenced the enormous potential of the non-premixed FTACLES method. The obtained results confirm the idea that SGS closure in diffusive combustion can be derived based on filtering arguments, and not only based on statistical approaches. The formalism offers an exact mathematical characterization of the modified flame structure resulting from a filtering operation for an unwrinkled flame front which cannot be resolved by the numerical grid. However numerous practical applications do not fulfill the condition of the SGS laminar flame front, and an additional wrinkling function should be considered. This introduces a statistical component, therefore the uncertainty due to the approximation performed by the PDF is to some extent shifted to the wrinkling expression. Further research might be carried out in the following directions:

- A conscientious investigation of the modified turbulence-chemistry interaction on a filtered flame front remains the biggest research issue. An efficiency function specifically designed for non-premixed combustion should be derived to deliver a fully consistent non-premixed FTACLES formulation. Machine learning strategies could be applied over targeted DNS data for the derivation, for which the resolved and SGS velocity and scalar dissipation appear as plausible variables. Subsequently the undergoing transformations and mechanisms should be substantiated from the theoretical point of view, correlations should be then derived to deliver a sound phenomenological interpretation.

- Regarding the sensor activation, a dynamic approach could be implemented to replace the constant value tolerance. The filtered flamelet numerical grid is in general finer than $\Delta/5$ in the region close to stoichiometry and a dynamic formulation with account for this discrepancy.

- A method limitation concerns the interpretation of the non-monotonically retrieved data, e.g. temperature and most of the species. This becomes particularly relevant for flame thickness lower than the filter size, where the profiles undergo a non-negligible trimming of the peak values, hence they cannot be directly compared with experimental statistics. A simplistic approach such as saving the unfiltered species data in terms of the filtered parameterizing variables, would deliver erroneous results for the flame structure relations do not hold anymore, and a reconstruction method should be developed.

- The performance of the NO transport equation approach in the current formalism should be evaluated. The pollutant prediction can be done in the adiabatic condition and then the model sensitivity to the addition of h as a fourth dimension to take into account radiation effect should be appraised. Concerning the transport equation closure, a consistent strategy should be proposed in order to link the filtered $\dot{\omega}_{NO}$ with the transport equation. The influence of c definition should be assessed and the consideration of NO in c definition might be justified considering its variation with K and therefore the progress variable's capability to adequately resolve this behavior.

- The inclusion of preferential diffusion effects within the formalism considering laminar non-premixed coflow flames and then turbulent diffusion conditions should be done. More complex flamelet configurations taking into account tangential effects might be employed to describe the laminar non-premixed coflow case, thus the non-counterflow effects will be undoubtedly separated from the filtering ones. Since the non-premixed filtered tabulated chemistry equations have been derived based on the counterflow flamelet configuration, the method correction terms should be recomputed for distinct flamelet governing equations. Targeting turbulent conditions, the effect of variable Le_i in counterflow flamelets as expressed within the FGM formulation can be included. For a (Z, c) parameterization, as the one employed in this study, a formulation modifying uniquely the c transport equation is recommended. The term representing the preferential diffusion, is tabulated in terms of the existing controlling parameters, hence it does not lead to any dimension increase.